English for Composite Materials and Engineering

复合材料与工程专业英语

第二版

郝凌云　主　编
陈晓玉　叶原丰　副主编

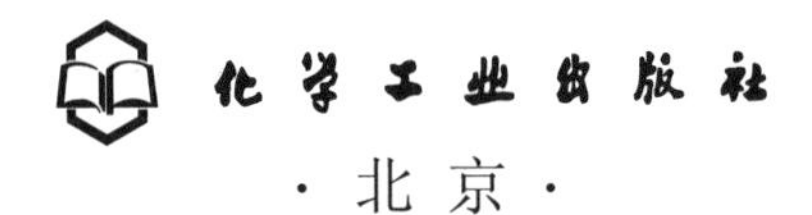

·北京·

内容提要

《复合材料与工程专业英语》旨在指导该专业学生及相关从业人员阅读本专业领域的英语书刊和文献，提高读者阅读英语科技资料的能力，并能进一步以英语为工具获取复合材料与工程专业所需要的信息。本书共分8章，共23个单元，其中第1章为复合材料概论，主要介绍复合材料的概念、分类及应用；第2章和第3章分别介绍基体材料和加强材料；第4章介绍界面的分类、力学特征和检测方法；第5章至第8章分别介绍高分子基复合材料、金属基复合材料、陶瓷基复合材料、纳米复合材料的概念、制造技术和应用。为提高学习兴趣，强化学习效果，在传统纸质教材的基础上，配套了互联网在线学习资源，包括：视频、原文朗读（音频、跟读）、在线习题、拓展阅读和PPT教学课件，以便师生使用。

本书除可作为普通高等院校复合材料与工程专业本科生教材外，还可作为相关材料专业的研究生、教师及企业技术人员在学习教学与研发过程的参考书。

图书在版编目（CIP）数据

复合材料与工程专业英语/郝凌云主编. —2版.—北京：化学工业出版社，2020. 7 (2023.1 重印)
ISBN 978-7-122-36920-8

Ⅰ.①复⋯ Ⅱ. ①郝⋯ Ⅲ. ①复合材料-英语-高等学校-教材 Ⅳ. ①TB33

中国版本图书馆CIP数据核字（2020）第081546号

责任编辑：王　婧　杨　菁　　装帧设计：王晓宇
责任校对：宋　夏

出版发行：化学工业出版社（北京市东城区青年湖南街13号　邮政编码100011）
印　　装：中煤（北京）印务有限公司
787mm×1092mm　1/16　印张 9¼　字数236千字　2023年1月北京第2版第3次印刷

购书咨询：010-64518888　　售后服务：010-64518899
网　　址：http: // www. cip. com. cn
凡购买本书，如有缺损质量问题，本社销售中心负责调换。

定　价：**49.00元**

前 言

复合材料具有鲜明的结构可设计性，通过先进材料制备技术可实现多组分性能互补和优化关联，对现代科学技术的发展有十分重要的作用，对复合材料研究的深度和应用的广度已成为衡量一个国家科学技术先进水平的重要标志之一。

随着材料科学与技术、工艺、应用的创新与发展，复合材料的发展对人才培养提出了迫切要求，迄今已有近五十所大学开设复合材料与工程本科专业。“复合材料与工程专业英语”作为综合专业知识与英语运用能力的综合课程，是学生国际化交流与应用能力的基石。

本教材编写基于“先进性、应用性、适教性”原则，参考复合材料与工程专业国际通用教材和学术研究刊物，内容涵盖金属基体、陶瓷基体、聚合物基体等复合材料的制备、界面控制、材料性能及应用等多个方面。本书可作为普通高等院校复合材料与工程专业本科生教材，也可供相关材料专业的研究生、教师及企业技术人员参考。

《复合材料与工程专业英语》第一版于 2014 年出版发行，自投入教学使用以来，得到广大师生的认可和好评。此次修订保留了第一版纸质教材的基本框架，并结合互联网的应用，增补了在线学习资源，包括视频、原文朗读（音频、跟读）、拓展阅读、在线习题、PPT 教学课件。所有线上资源可以通过扫描纸质教材相关章节处的二维码快速获取。由 JEC 集团授权使用的英文视频“I'm a composite material”，对复合材料与工程专业学习起到提纲挈领、一览全局的作用。

本书由郝凌云主编，陈晓玉统稿。其中第 1 章由郝凌云、张伟编写；第 2 章由叶原丰编写；第 3 章由林青编写；第 4 章由陈晓玉、李俊琳编写；第 5 章由陈晓宇编写；第 6 章由张小娟、赵媛编写；第 7 章由韦鹏飞编写；第 8 章由张小娟和赵媛编写；教材后的阅读理解答案和词汇表由张伟统一编撰、汇总。线上资源的原文朗读音频由宗小琦、黄燕录制完成。

特别致谢江苏省复合材料学会、金陵科技学院外国语学院在本次修订过程中给予的指导与帮助。

由于编者水平有限，本书难免有疏漏之处，敬请广大读者批评指正。

教学课件

编者

2020 年 1 月于南京

第一版前言

专业英语是从语言学习到信息交流的发展，也是大学英语教学中一个不可缺少的重要的组成部分，担负着促使学生学会在专业领域中用英语去进行有实际意义的交流。高质量的专业英语教材是完成专业英语教学的基础。

当今社会与外语联系越来越紧密，同时也对复合材料与工程专业的学生掌握英语的程度有了更高的要求。《复合材料与工程专业英语》是根据大学英语教学大纲（理工科本科用）的专业阅读部分的要求编写的。编写的主要目的是扩充学生的复合材料与工程专业的词汇量，提高学生阅读和翻译英语文献和资料的能力，深化学生对本专业关键技术的认识，了解本学科目前的进展与动向，从而契合该专业的工程化教育及学生的国际化培养。

本教材收集了陶瓷基体、聚合物基体、金属基体等复合材料的制备、界面控制、材料性能及应用等复合材料与工程专业领域的最新英语文献，共有八章内容，每章包括精读和泛读两部分，均设有词汇表、注释和练习项目，读者可在掌握复合材料与工程专业英语的翻译及写作技巧的同时进一步学习专业的有关知识。

本教材由郝凌云主编。其中第 1 章由郝凌云、冯志强编写；第 2 章由叶原丰编写；第 3 章由林青编写；第 4 章由陈晓玉、李俊琳编写；第 5 章由陈晓宇编写；第 6 章和第 8 章由张小娟编写；第 7 章由韦鹏飞编写；教材后的阅读理解和词汇表由陈晓玉统一编撰、汇总。

由于编者水平有限，对本书中不当之处，敬请广大读者批评指正。

郝凌云
2014 年 1 月

CONTENTS

Chapter 1 Introduction to composite materials

扫码看视频 扫码听音频

Unit 1 Introduction and conception

Natural and Man-made Composites

A composite is a material that is formed by combining two or more materials to achieve some superior properties. Almost all the materials which we see around us are composites. Some of them like woods, bones, stones, etc. are natural composites, as they are either grown in nature or developed by natural processes. Wood is a fibrous material consisting of thread-like hollow elongated organic cellulose that normally constitutes about 60%-70% of wood of which approximately 30%-40% is crystalline, insoluble in water, and the rest is amorphous and soluble in water. Cellulose fibres are flexible but possess high strength. The more closely packed cellulose provides higher density and higher strength. The walls of these hollow elongated cells are the primary load-bearing components of trees and plants. When the trees and plants are live, the load acting on a particular portion (e.g., a branch) directly influences the growth of cellulose in the cell walls located there and thereby reinforces that part of the branch, which experiences more forces. This self-strengthening mechanism is something unique that can also be observed in the case of live bones. Bones contain short and soft collagen fibres i.e., inorganic calcium carbonate fibres dispersed in a mineral matrix called apatite. The fibres usually grow and get oriented in the direction of load. Human and animal skeletons are the basic structural frameworks that support various types of static and dynamic loads. Tooth is a special type of bone consisting of a flexible core and the hard enamel surface. The compressive strength of tooth varies through the thickness. The outer enamel is the strongest with ultimate compressive strength as high as 700MPa. Tooth seems to have piezoelectric properties i.e., reinforcing cells are formed with the application of pressure. The most remarkable features of woods and bones are that the low density, strong and stiff fibres are embedded in a low density matrix resulting in a strong, stiff and lightweight composite (Tab. 1.1). It is therefore no wonder that early development of aero-planes should make use of woods as one of the primary structural materials, and about two hundred million years ago, huge flying amphibians, pteranodons and pterosaurs, with wing spans of 8-15 m, could soar from the mountains like the present-day hang-gliders. Woods and bones in many respect, may be considered to be predecessors to modern man-made composites.

Tab. 1.1 Typical mechanical properties of natural fibres and natural composites

Materials	Density /(kg/m³)	Tensile modulus /GPa	Tensile strength /MPa
Fibres			
Cotton	1540	1.1	400
Flax	1550	i	780
Jute	850	35	600
Coir	1150	4	200
Pineapple leaf	1440	65	1200
Sisal	810	46	700
Banana	1350	15	650
Asbestos	3200	186	5860
Composites			
Bone	1870	28	140
Ivory	1850	17.5	220
Balsa	130	3.5	24
Spruce	470	11	90
Birch	650	16.5	137
Oak	690	13	90
Bamboo	900	20.6	193

Early men used rocks, woods and bones effectively in their struggle for existence against natural and various kinds of other forces. The primitive people utilized these materials to make weapons, tools and many utility-articles and also to build shelters. In the early stages they mainly utilized these materials in their original form. They gradually learnt to use them in a more efficient way by cutting and shaping them to more useful forms. Later on they utilized several other materials such as vegetable fibres, shells, clays as well as horns, teeth, skins and sinews of animals.

Woods, stones and clays formed the primary structural materials for building shelters. Natural fibres like straws from grass plants and fibrous leaves were used as roofing materials. Stone axes, daggers, spears with wooden handles, wooden bows, fishing nets woven with vegetable fibers, jewelleries and decorative articles made out of horns, bones, teeth, semiprecious stones, minerals, etc. were but a few examples that illustrate how mankind, in early days, made use of those materials. The limitations experienced in using these materials led to search for better materials to obtain a more efficient material with better properties. This, in turn, laid the foundation for development of man-made composite materials.

The most striking example of an early man-made composite is the straw-reinforced clay which molded the civilization since prehistoric times. Egyptians, several hundred years B.C., were known to reinforce the clay like deposits of the Nile Valley with grass plant fibres to make sun baked mud bricks that were used in making temple walls, tombs and houses. The watch towers of the far western Great Wall of China were supposed to have been built with straw-reinforced bricks during the Han Dynasty (about 200 years B.C.). The natural fibre reinforced clay, even today continues to be one of the primary housing materials in the rural sectors of many third world countries.

The other classic examples are the laminated wood furniture used by early Egyptians (1500 B.C.), in which high quality wood veneers are bonded to the surfaces of cheaper woods. The origin of paper which made use of plant fibres can be traced back to China (108 A.D.). The bows used by the warriors under the Mongolian Chief Djingiz Chan (1200 A.D.) were believed to be made with the adhesive bonded laminated composite consisting of buffalo or antilope horns, wood, silk and ox-neck tendons. These laminated composite bows could deliver arrows with an effective shoot in range of about 740 m.

Potteries and hydraulic cement mortars are some of the earliest examples of ceramic composites. The cloisonne ware of ancient China is also a striking example of wire reinforced ceramics. Fine metallic wires were first shaped into attractive designs which were then covered with colored clays and baked. In subsequent years, fine metallic wires of various types were cast with different metal and ceramic matrices and were utilized in diverse applications. Several other matrix materials such as natural gums and resins, rubbers, bitumen, shellac, etc. were also popular. Naturally occurring fibres such as those from plants (cotton, flax, hemp, etc.), animals (wool, fur and silk) and minerals (asbestos) were in much demand. The high value textiles woven with fine gold and silver threads received the patronage from the royalty and the rich all over the world. The intricate, artful gold thread embroidery reached its zenith during the Mughal period in the Indian subcontinent. The glass fibres were manufactured more than 2000 years ago in Rome and Mesopotamia and were abundantly used in decoration of flower vases and glass wares in those days.

The twentieth century has noticed the birth and proliferation of a whole gamut of new materials that have further consolidated the foundation of modern composites. Numerous synthetic resins, metallic alloys and ceramic matrices with superior physical, thermal and mechanical properties have been developed. Fibres of very small diameter (<10 μm) have been drawn from almost all materials. They are much stronger and stiffer than the same material in bulk form. The strength and stiffness properties have been found to increase dramatically, when whiskers (i.e., single crystal fibers) are grown from some of these materials. Fig. 1.1 illustrates the specific tensile strength and the specific tensile modulus properties are obtained by dividing the strength (MPa) and modulus (GPa) by either the density (kg/m^3) or the specific gravity of the material. Because of the superior mechanical properties of fibers, the use of fibers as reinforcements started gaining momentum during the twentieth century. The aerospace industries took the lead in using fiber reinforced laminated plastics to replace several metallic parts. The fibres like glass, carbon, boron and Kevlar, and plastics such as phenolics, epoxies and polyesters caught the imagination of composite designers. One major advantage of using fibre reinforced plastics (FRP) instead of metals is that they invariably lead to a weight efficient design in view of their higher specific modulus and strength properties.

Composites, due to their heterogeneous composition, provide unlimited possibilities of deriving any characteristic material behavior. This unique flexibility in design tailoring plus other attributes like ease of manufacturing, especially molding to any shape with polymer composites, repairability, corrosion resistance, durability, adaptability, cost effectiveness, etc. have attracted the attention of many users in several engineering and other disciplines. Every industry is now vying with each other to make the best use of composites. One can now notice the application of composites in many disciplines starting from sports goods to space vehicles. This worldwide interest during the last four decades has led to the prolific advancement in the field of composite materials and structures. Several high performance polymers have now been developed. Substantial progress has been made in the development of stronger and stiffer fibres, metal and ceramic matrix composites, manufacturing and machining processes, quality control and nondestructive evaluation techniques, test methods as well as design and analysis methodology. The modern man-made composites have now firmly established as the future material and are destined to dominate the material scenario right through the twenty-first century.

[*Copy from* Sinha P K. Composite materials and structures[M]. India Kharagpur: Composite Centre of Excellence, AR & DB, Department of Aerospace Engineering, I. I. T. Kharagpur, 2006.]

New words and expressions

amorphous *adj*. 无组织的，模糊的；无固定形状的，非结晶的
load-bearing 支撑结构
elongated cell 细长细胞
collagen fibre 胶原纤维
apatite *n*. 磷灰石
enamel surface 釉质表面
piezoelectric property 压电性能
amphibian *n*. 两栖动物，水旱两生植物；水陆两用车，水陆两用飞行器
adj. 两栖（类）的，水陆两用的；具有双重性的
pteranodon *n*. 无齿翼龙
pterosaur *n*. 翼龙
hang-glider 滑翔机
flax *n*. 亚麻
jute *n*. 黄麻
coir *n*. 椰子壳的纤维；棕
sisal *n*. 剑麻
balsa *n*. 西印度白塞木
spruce *n*. 云杉
birch *n*. 桦木
veneer *n*. 表层饰板
bow *n*. 箭弓
pottery *n*. 陶器
hydraulic cement mortar 液压水泥砂浆
cloisonne ware 景泰蓝制品
bitumen *n*. 沥青
shellac *n*. 虫胶
gold thread embroidery 金线刺绣
phenolics *n*. 酚醛塑料
epoxy *n*. 环氧树脂
polyester *n*. 聚酯，涤纶
heterogeneous *adj*. 多种多样的；不均匀的，异质的

Notes

(1) When the trees and plants are live, the load acting on a particular portion (e.g., a branch) directly influences the growth of cellulose in the cell walls located there and thereby reinforces that part of the branch, which experiences more forces.
存活树木或植物的纤维素在其特殊部位（如枝干）的附着支撑作用会直接影响到该部位细胞壁上纤维素的生长，故使枝干等部位表现出更耐受强力的特征。

(2) The most remarkable features of woods and bones are that the low density, strong and stiff fibres are embedded in a low density matrix resulting in a strong, stiff and lightweight composite.
木材和骨头具有密度低、强度高和柔韧好的显著特点，因为纤维沉积在低密度的介质中会形成高强度、高韧性和轻质地的复合材料。

(3) Substantial progress has been made in the development of stronger and stiffer fibres, metal and ceramic matrix composites, manufacturing and machining processes, quality control and nondestructive evaluation techniques, test methods as well as design and analysis methodology.
寻找更强、更韧纤维、金属和陶瓷基复合材料，优化复合材料制造及加工工艺、探究质量控制和无损伤检测技术以及对其设计和评估等复合材料领域的各方面研究都在持续进展。

Exercises

1. Question for discussion

(1) The modern man-made composites have now firmly established as the future material and are destined to dominate the material scenario right through the twenty-first century. What are the main reasons about this trend?

(2) Why did air-planes use wood as one of the primary structural materials in the early days?

(3) Are there any natural composites in life that are not covered in this article?

2. Translate the following into Chinese

(1) Tooth is a special type of bone consisting of a flexible core and the hard enamel surface. The compressive strength of tooth varies through the thickness. The outer enamel is the strongest with ultimate compressive strength as high as 700MPa.

(2) It is therefore no wonder that early development of aero-planes should make use of woods as one of the primary structural materials, and about two hundred million years ago, huge flying amphibians, pterendons and pterosaurs, with wing spans of 8-15 m, could soar from the mountains like the present-day hang-gliders.

(3) Early men used rocks, woods and bones effectively in their struggle for existence against natural and various kinds of other forces. The primitive people utilized these materials to make weapons, tools and many utility-articles and also to build shelters.

(4) Stone axes, daggers, spears with wooden handles, wooden bows, fishing nets woven with vegetable fibers, jewelleries and decorative articles made out of horns, bones, teeth, semiprecious stones, minerals, etc. were but a few examples that illustrate how mankind, in early days, made use of those materials.

(5) The twentieth century has noticed the birth and proliferation of a whole gamut of new materials that have further consolidated the foundation of modern composites.

(6) Substantial progress has been made in the development of stronger and stiffer fibers, metal and ceramic matrix composites, manufacturing and machining processes, quality control and non-destructive evaluation techniques, test methods as well as design and analysis methodology.

3. Translate the following into English

(1) 我们身边看到所有的材料几乎都是复合材料。

(2) 纤维素中空细长细胞是树和其它植物结构的基本支撑组分。

(3) 许多观点认为，木头和骨头是现代人造复合材料的前身。

(4) 即使在今天，天然纤维增强粘土仍然是许多第三世界国家农村区域的主要建房材料之一。

(5) 向复合材料中增加晶须（即，单晶纤维）后发现材料的强度和韧性显著增加了。

(6) 陶瓷和水泥砂浆是最早的陶瓷复合材料案例。

(7) 航空业率先使用纤维增强基层状塑料材质代替其部分金属部件。

(8) 由于复合材料其不同成分，因此获得任何特征材料特性的各种可能性。

(9) 现在各个行业通过使用更好的复合材料进行竞争。

4. Scenario simulation

As a college student majoring in composite materials, please write a job application letter to introduce your major and use no more than 200 words.

Unit 2 Classifications and terminology of composite materials

Composite materials are usually classified according to the type of reinforcement used. Two broad classes of composites are fibrous and particulates. Each has unique properties and application potential, and can be subdivided into specific categories as discussed below.

1. Fibrous

A fibrous composite consists of either continuous (long) or chopped (whiskers) fibers suspended in a matrix material. Both continuous fibers and whiskers can be identified from a geometric viewpoint.

(1) Continuous Fibers A continuous fiber is geometrically characterized as having a very high length-to-diameter ratio. They are generally stronger and stiffer than bulk material. Fiber diameters generally range between 0.00012 and 0.0074 in. (3-200 μm) depending upon the fiber.

(2) Whiskers A whisker is generally considered to be a short, stubby fiber. It can be broadly defined as having a length-to-diameter ratio of $5 < l/d < 1000$ and beyond. Whisker diameters generally range between 0.787 and 3937 μin. (0.02-100 μm).

Composites in which the reinforcements are discontinuous fibers or whiskers can be produced so that the reinforcements have either random or biased orientation. Material systems composed of discontinuous reinforcements are considered single layer composites. The discontinuities can produce a material response that is anisotropic, but in many instances the random reinforcements produce nearly isotropic composites.

Continuous fiber composites can be either single layer or multilayered. The single layer continuous fiber composites can be either unidirectional or woven, and multilayered composites are generally referred to as laminates. The material response of a continuous fiber composite is generally orthotropic. Schematics of both types of fibrous composites are shown in Fig. 1.1.

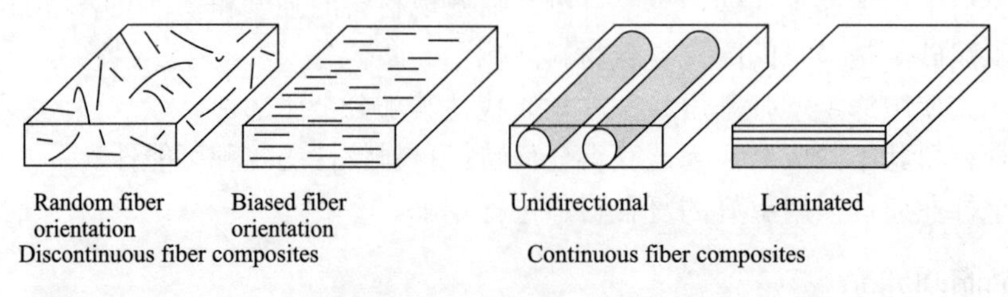

Fig. 1.1 Schematic representation of fibrous composites

2. Particulate

A particulate composite is characterized as being composed of particles suspended in a matrix. Particles can have virtually any shape, size or configuration. Examples of well-known particulate composites are concrete and particle board. There are two subclasses of particulates: flake and filled/skeletal.

(1) Flake A flake composite is generally composed of flakes with large ratios of platform area to thickness, suspended in a matrix material (particle board, for example).

(2) Filled/Skeletal A filled skeletal composite is composed of a continuous skeletal matrix filled by a second material: for example, a honeycomb core filled with an insulating material.

The response of a particulate composite can be either anisotropic or orthotropic. Such composites are used for many applications in which strength is not a significant component of the design. A schematic of several types of particulate composites is shown in Fig. 1.2.

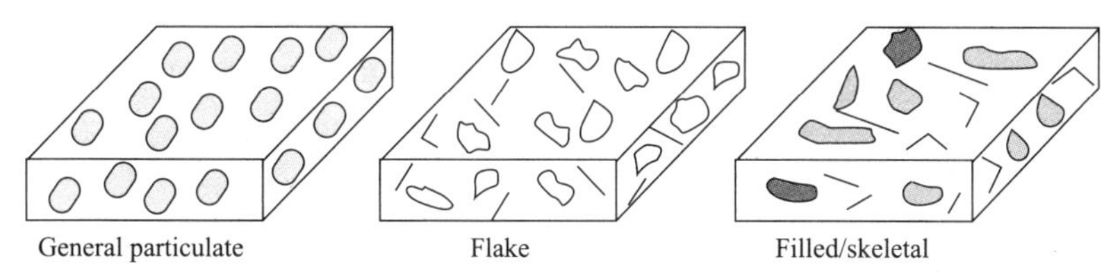

Fig. 1.2 Schematic representation of particulate composites

3. Fundamental Terminology

Some of the more prominent terms used with composite materials are defined below. A more detailed list can be found in Tsai, as well as in the Glossary.

(1) Lamina A lamina is a flat (or sometimes curved) arrangement of unidirectional (or woven) fibers suspended in a matrix material. A lamina is generally assumed to be orthotropic, and its thickness depends on the material from which it is made. For example, a graphite/epoxy (graphite fibers suspended in an epoxy matrix) lamina may be on the order of 0.005 in. (0.127 mm) thick. For the purpose of analysis, a lamina is typically modeled as having one layer of fibers through the thickness. This is only a model and not a true representation of fiber arrangement. Both unidirectional lamina and woven lamina are schematically shown in Fig. 1.3.

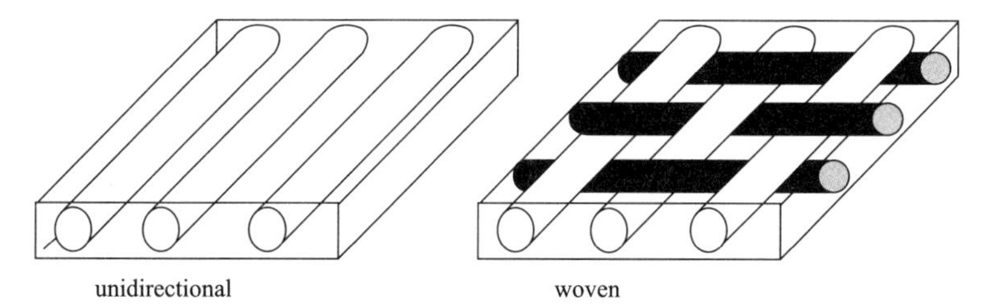

Fig. 1.3 Schematic representation of unidirectional and woven composite lamina

(2) Reinforcements Reinforcements are used to make the composite structure or component stronger. The most commonly used reinforcements are boron, glass, graphite (often referred to as simply carbon), and Kevlar, but there are other types of reinforcements such as alumina, aluminum, silicon carbide, silicon nitride, and titanium.

(3) Fibers Fibers are a special case of reinforcements. They are generally continuous and have diameters ranging from 120 to 7400 μin. (3-200 μm). Fibers are typically linear elastic or elastic-perfectly plastic and are generally stronger and stiffer than the same material in bulk form. The most commonly used fibers are boron, glass, carbon, and Kevlar. Fiber and whisker technology is continuously changing.

(4) Matrix The matrix is the binder material that supports, separates, and protects the fibers. It provides a path by which load is both transferred to the fibers and redistributed among the fibers in the event of fiber breakage. The matrix typically has a lower density, stiffness, and strength than the fibers. Matrices can be brittle, ductile, elastic, or plastic. They can have either linear or nonlinear stress-strain behavior. In addition, the matrix material must be capable of being forced around the reinforcement during some stage in the manufacture of the composite. Fibers must often be chemically treated to ensure proper adhesion to the matrix. The most commonly used matrices are carbon, ceramic, glass, metal, and polymeric. Each has special appeal and usefulness, as well as limitations. Richardson presents a comprehensive discussion of matrices, which guided the following presentation.

① **Carbon Matrix** A carbon matrix has a high heat capacity per unit weight. They have been used as rocket nozzles, ablative shields for reentry vehicles, and clutch and brake pads for aircraft.

② **Ceramic Matrix** A ceramic matrix is usually brittle. Carbon, ceramic, metal, and glass fibers are typically used with ceramic matrices in areas where extreme environments (high temperatures, etc.) are anticipated.

③ **Glass Matrix** Glass and glass-ceramic composites usually have an elastic modulus much lower than that of the reinforcement. Carbon and metal oxide fibers are the most common reinforcements with glass matrix composites. The best characteristics of glass or ceramic matrix composites is their strength at high service temperatures. The primary applications of glass matrix composites are for heat-resistant parts in engines, exhaust systems, and electrical components.

④ **Metal Matrix** A metal matrix is especially good for high-temperature use in oxidizing environments. The most commonly used metals are iron, nickel, tungsten, titanium, magnesium, and aluminum. There are three classes of metal matrix composites:

a. Class Ⅰ The reinforcement and matrix are insoluble (there is little chance that degradation will affect service life of the part). Reinforcement/matrix combinations in this class include tungsten or alumina/copper, BN-coated B or boron/aluminum, and boron/magnesium.

b. Class Ⅱ The reinforcement/matrix exhibits some solubility (generally over a period of time and during processing) and the interaction will alter the physical properties of the composite. Reinforcement matrix combinations included in this class are carbon or tungsten/nickel, tungsten/columbium, and tungsten/copper (chromium).

c. Class Ⅲ The most critical situations in terms of matrix and reinforcement are in this class. The problems encountered here are generally of a manufacturing nature and can be solved through processing controls. Within this class the reinforcement matrix combinations include alumina or boron or silicon carbide/titanium, carbon or silica/aluminum, and tungsten/copper (titanium).

⑤ **Polymer Matrix** Polymeric matrices are the most common and least expensive. They are found in nature as amber, pitch, and resin. Some of the earliest composites were layers of fiber, cloth, and pitch. Polymers are easy to process; offer good mechanical properties, generally wet reinforcements well, and provide good adhesion. They are a low-density material. Because of low processing temperatures, many organic reinforcements can be used. A typical polymeric matrix is either viscoelastic or viscoplastic, meaning it is affected by time, temperature, and moisture. The terms *thermoset* and *thermoplastic* are often used to identify a special property of many polymeric matrices.

a. Thermoplastic. A thermoplastic matrix has polymer chains that are not cross-linked.

Although the chains can be in contact, they are not linked to each other. A thermoplastic can be remolded to a new shape when it is heated to approximately the same temperature at which it was formed.

b. Thermoset. A thermoset matrix has highly cross-linked polymer chains. A thermoset can not be remolded after it has been processed. Thermoset matrices are sometimes used at higher temperatures for composite applications.

(5) Laminate A laminate is a stack of lamina, as illustrated in Fig. 1.4, oriented in a specific manner to achieve a desired result. Individual lamina is bonded together by a curing procedure that depends on the material system used. The mechanical response of a laminate is different from that of the individual lamina that forms it. The laminate's response depends on the properties of each lamina, as well as the order in which the laminae are stacked.

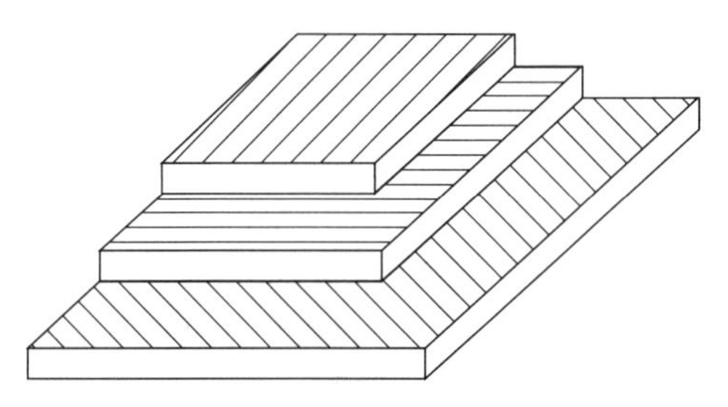

Fig. 1.4 Schematic of a laminated composite

(6) Micromechanics A specialized area of composites involving a study of the interaction of constituent materials on the microscopic level. This study is generally conducted by use of a mathematical model describing the response of each constituent material.

(7) Macromechanics A study of the overall response of a lamina (or laminate) in which the effects of constituent materials are averaged to achieve an apparent response on the macroscopic level.

[*Selected from* Tsai S W, Hahn S W. Introduction to composite materials[M].Technomic Publishing Co, 1980.]

New words and expressions

geometric *adj.* 几何学的，几何装饰的，成几何级数增减的
unidirectional *adj.* 单向的，单向性的
woven *n.* 交叉织状
v. 编，织，织成（weave 的过去分词）
orthotropic *adj.* [力]正交各向异性；[植]直生的，正交的
lamina *n.* 薄板，薄层；名词复数 laminae，laminas
reinforcement *n.* 增强体
boron *n.* 硼砂
ductile *n.* 韧性
thermoplastic *n.* 热塑性
thermoset *n.* 热固性
micromechanic *n.* 微观机理
macromechanic *n.* 宏观机理

Notes

(1) is generally considered to be 通常意义上

(2) be subdivided into 细分为

(3) biased orientation 偏置取向

(4) Fundamental Terminology 基本术语

(5) elastic perfectly plastic 理想弹塑性

Exercises

1. Question for discussion

(1) What are the two broad classes of composites? Simply describe the unique properties and potential applications of each one.

(2) What is the general method used to study the Micromechanics of composites?

2. Translate the following into Chinese

(1) The discontinuities can produce a material response that is anisotropic, but in many instances the random reinforcements produce nearly isotropic composites.

(2) The single layer continuous fiber composites can be either unidirectional or woven, and multi-layered composites are generally referred to as laminates.

(3) The response of a particulate composite can be either anisotropic or orthotropic. Such composites are used for many applications in which strength is not a significant component of the design.

(4) The most commonly used reinforcements are boron, glass, graphite (often referred to as simply carbon), and Kevlar, but there are other types of reinforcements such as alumina, aluminum, silicon carbide, silicon nitride, and titanium.

(5) The most commonly used fibers are boron, glass, carbon, and Kevlar. Fiber and whisker technology is continuously changing.

(6) Micromechanics. A specialized area of composites involving a study of the interaction of constituent materials on the microscopic level. This study is generally conducted by use of a mathematical model describing the response of each constituent material.

(7) Macromechanics. A study of the overall response of a lamina (or laminate) in which the effects of

constituent materials are averaged to achieve an apparent response on the macroscopic level.

(8) The mechanical response of a laminate is different from that of the individual lamina that forms it. The laminate's response depends on the properties of each lamina, as well as the order in which the lamina are stacked.

3.Translate the following into English

(1) 复合材料分类通常是依据所使用增强剂组分的不同进行的。

(2) 连续纤维的几何特征是具有很高的长径比。

(3) 连续纤维增强体复合材料的结构既可以是单层的也可以是多层的。

(4) 复合材料中几种粒子形态如图 1.2 所示。

(5) 常用热固性和热塑性来衡量许多聚合物基体的性质。

(6) 纤维通常意义上被看作线弹性或理想弹性材料，一般比同类块体材料更强、更韧。

(7) 玻璃和玻璃陶瓷复合材料的弹性模量通常比钢筋的弹性模量低很多。

(8) 玻璃基复合材料主要应用于发动机、排气系统和电器元件的耐热部件。

(9) 这里遇到的问题通常属于制造性质，可以通过处理控制来解决。

4. Scenario simulation

You are a science popularizing worker, in the weekend you will introduce to the children what is the composite material, introduce the classification of composite materials, and illustrate the common composite materials belonging to which category.

Write a report that children can understand easily and no more than 200 words.

在线习题

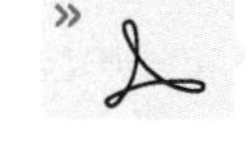
拓展阅读

Unit 3 Applications

1. Introduction

Polymer composites have been widely used for several years and their market share is continuously growing. It is widely known that the use of a polymer and one (or more) solid fillers allows obtaining several advantages and, in particular, a combination of the main properties of the two (or more) solid phases. Among the fillers used, it is worth citing calcium carbonate, glass fibers, talc, kaolin, mica, wollastonite, silica, graphite, synthetic fillers (e.g. PET- or PVA-based fibers), high-performance fibers (carbon, aramidic, etc.). However, this leads also to one of the main limitations of polymer composites: the two different components make the reuse and recycling quite difficult, to such an extent that it is often preferred to perform the direct disposal in a dump, or incineration. This way is often considered to be unsatisfactory (especially in the first case), because of the high costs, the technical difficulties and the environmental impact. The latter is, indeed, a problem of primary importance. Furthermore, it is worsened by the fact that plastics production requires a remarkable consumption of oil-based resources, which are notoriously non-renewable. These problems have begun to be particularly evident for about 10 years, thus leading the scientific research to look for new alternatives, able to replace traditional polymer composites with substitutes having lower environmental impact and thus often referred to as "ecocomposites" or "green composites". This task can be made easier by the fact that many of the typical applicative fields of these composites do not require excellent mechanical properties (i.e. secondary and tertiary structures, panels, packaging, gardening items, cases, etc.)

In chronological order, the first attempts in this direction were focused on the production and characterization of polymer composites based on recyclable polymers (e.g. polyolefins) filled with natural-organic fillers, i.e. fibers and particles extracted from plants. Several points support this choice; first of all, the use of natural-organic fillers in replacement for traditional mineral inorganic ones allows a considerable reduction in the use of nonbiodegradable polymers and non-renewable resources. Furthermore, these fillers are usually drawn from relatively abundant plants (often from wastes), therefore they are very cheap. They are also much less abrasive than inorganic-mineral counterparts to processing machinery, less dangerous for the production employees in case of inhalation, easy to be incinerated, they lead to final composites with lower specific weight (in comparison to mineral-filled counterparts) and allow obtaining interesting properties in terms of thermal and acoustic insulation.

The most widely known and used natural-organic fillers are wood flour and fibers. Wood flour can be easily and cheaply obtained from sawmill wastes and it is usually used after proper sieving. Wood fibers are produced by thermo-mechanical processes on wood waste.

Besides wood derivatives, other natural-organic fillers have begun to find application as well. Among these, some examples are cellulose, cotton, flax, sisal, kenaf, jute, hemp, starch. Further "environment-friendliness" can be achieved upon using post-consumer recycled plastics in place of virgin polymer matrices.

Wood flour and fibers are quite interesting because of the low cost, dimensional stability, elastic modulus, while tensile properties do not improve; the main shortcomings are the poor adhesion between the filler particles and the polymer matrix, low impact strength, thermal decomposition at temperatures over 200℃.

Flax, sisal, hemp and kenaf are relatively similar and are basically long fibers extracted from the bast of the plants; they can be used as fillers by proper cutting into long or short fibers. Starch is a polysaccharide present in many plants acting as an energy reservoir. It is made of glucose monomers linked by α- (1-4) bonds. In general, the addition of granular starch to a polymer leads to a reduction of the elongation at break (and often of the tensile stress as well), as high as the starch content increases, while the elastic modulus is enhanced. A limitation of this filler type is the tendency to absorb water because of its very high surface area and its hygroscopic nature.

Other less used natural-organic fillers can include rice husk ash, nutshells, oil palm empty fruit bunch fibers, corn plants extracted fibers, etc.

2. Industrial applications

With concern to the industrial applications, several paths have been undertaken. In short, it can be stated that the most used natural-organic filler is wood (either flour or fibers), especially as low cost filler for polyolefins. Wood flour is usually obtained from sawmill waste after a simple sieving treatment; wood fibers are produced from sawmill waste by a wet thermomechanical process. Already explored industrial applications include window and doorframes, furniture, railroad sleepers, automotive panels and upholstery, gardening items, packaging, shelves and in general those applications which do not require very high mechanical resistance but, instead, low purchasing and maintenance costs. Furthermore, it is possible and convenient to use recycled polymers in place of virgin ones, thus assuring improved cost-efficiency and eco-sustainability. Some examples of industrial applications can be easily found on the technical literature and on the Internet; these include, for instance, indoor furniture panels, footboards and platforms, automotive panels and upholstery, noise insulating panels, etc., mainly produced by American, German, Japanese, British and Italian firms.

In particular, the role of automotive industry in this field is of primary importance. The first carmaker to use polymers filled with natural fibers was Mercedes-Benz in the 90s, by manufacturing door panels containing jute fibers. This "example" was soon followed by other main carmakers, which have started to utilize polymer composites with natural-organic fillers as materials for door panels, roof upholstery, headrests, parcel shelves, etc., thanks to the advantages they can assure in terms of environmental impact, weight, elastic modulus and costs. Depending on the applications, it was sometimes necessary to improve the mechanical properties through fiber pre-treatment (acetylation, use of MAgPP, etc.), and the treated fibers were then used in several ways, in order obtaining mats, non-woven structures, etc.

Some authors assert also that, by means of special treatments on natural fibers, these could lead to the production of high-quality composites with mechanical properties comparable to glass fiber filled composites. A result which would be impossible to obtain otherwise, since the hydrophilic nature of natural-organic fillers favors agglomeration, humidity absorption and lack of adhesion with the polymer matrix; in fact, many efforts are being done to overcome the problem of interfacial adhesion. Among these, the investigations on silane-based adhesion promoters or some fiber treatments with alkaline substances or dilute resins. However, a proper use of long (i.e. more than approx. 5 cm, at least) fibers can already allow obtaining composites suitable to semi-structural applications.

3. Towards complete environmental sustainability

Some recent developments are pushed by the fact that, unfortunately, even these "green" composites are not fully ecocompatible, since their recyclability has some limitations (temperatures can not exceed about 200℃ during recycling and all the main properties will worsen because of the degradation phenomena) and their biodegradability regards only the filler but not, of course, traditional (petroleum-based) polymer matrices. For this reason, research developments during the last years have been focused on the production on 100% eco-sustainable and "green" composites, by replacing non-biodegradable polymer matrices with biodegradable ones. Several biodegradable, natural-derived polymers exist, such as polysaccharides (starch, chitin, collagen, gelatines, etc.), proteins (casein, albumin, silk, elastin, etc.), polyesters [e.g. poly(hydroxyalcanoate), poly (hydroxybutyrate), polylactic acid)], lignin, lipids, natural rubber, some polyamides, polyvinyl alcohols, polyvinyl acetates, and polycaprolactone. In the majority of cases, they degrade through enzymatic reactions in suitable environments (typically, humid).

As a general rule, biodegradable polymers can be classified according to their origin, i.e. into agropolymers (e.g. starch), microbial-derived (e.g. PHA) and chemically synthesized from agro-based monomers (e.g. PLA) or conventional monomers (e.g. synthetic polyesters). Several examples are available in literature. For instance, Japanese researchers have investigated on composites based on starch and bamboo fibers. Others have studied the interesting properties of composites based on Monsanto Biopol (a polyhydroxybutyrate-hydroxyvalerate copolymer) with ananas and jute fibers. In some cases, natural fibers were pre-treated by alkali treatment and chemical groups grafting.

Work has also been done to use matrices based on soy proteins. For instance, Netravali and coworkers used soy proteins in combination with several natural fibers, obtaining interesting composites which, in some cases, show global characteristics even superior to many wood types. An interesting system for automotive applications is a composite where the polymer matrix is based on soy and corn oil, which are used as raw materials for the production of the polymer (in a way which is comparable with the one by which commodity polymers come from oil) with good resistance, flexibility, lightness and durability properties. Examples of possible applications are panels, seats, packaging, furnishing, etc. Some companies have been working on synthetic silk production by genetic engineering, which may be used for biodegradable materials as well.

A very interesting class of biodegradable polymers which could be used for eco-composites production is the Novamont's Mater-Bi. It is known that Mater-Bi family polymers are usually based on modified starch and synthetic polymers (polyesters inprevalence) and are compostable. Literature reports that Mater-Bi can be conveniently used for the realization of fully biodegradable composites, however mechanical properties, processability and biodegradability strongly depend on the Mater-Bi and the filler used, filler content, processing techniques. For instance, rigidity of the materials usually increases upon increasing filler content, while ultimate properties can change significantly depending on filler content, processing techniques (mixing, extrusion, injection molding, etc.), processing parameters (speed, temperature, etc.). Scaffaro et al. found that injection molding can significantly reduce the polymer's molecular weight in comparison to extrusion followed by compression molding but, on the other hand, it can improve both surface and internal morphology of the composite samples, thus leading to significant increases of tensile strength if compared to the neat, unfilled samples.

An example of the way how the main mechanical properties can change upon going from a neat Mater-Bi grade to the corresponding, 30 wt.% filled green composites, is provided in Tab. 1.2. The composites were filled with coarse (SDc) and fine (SDf) wood flour. The filler was always dried before processing, while the polymer was dried only in the case of “dry” -labeled samples.

Tab. 1.2 Main mechanical properties for a neat Mater-Bi grade("MB") and the corresponding 30 wt.% wood flour(SDc=coarse, SDf=fine) filled green composites, without ("humid") and with ("dry") drying treatment on the polymer before processing

Property	MB	MB+30%SDc (humid)	MB+30%SDc (humid)	MB+30%SDc (dry)	MB+30%SDc (dry)
Elastic modulus/MPa	88	457	483	442	530
Tensile strength/MPa	6	5.5	6.4	7	6.7
Elongation at break/%	73	2.8	2.7	2.3	2.3
Heat deflection temperature/℃	39	49	54	48	55
Impact strength/(J/m)	No break	86	54	63	44

It can be observed that wood flour significantly enhances the stiffness and the thermomechanical resistance of the materials; furthermore, tensile strength keeps practically constant, probably thanks to the hydrophilic nature of the chemical groups present in the Mater-Bi matrix; only the ductility is worsened. However, since polyesters are typically present in the composition of most Mater-Bi grades, a drying pre-treatment is advisable in order reducing the hydrolytic chain-scission reactions due to the presence of water in the system during processing.

Biodegradability of Mater-Bi based green composites has not been much investigated so far. The papers available regard biodegradation of composites in soil and found that the composites underwent biodegradation after burial in soil. Rutkowska et al. studied biodegradation in different natural environments, finding a complete biodegradation after 4 weeks. However, these investigations regarded just neat materials.

Scaffaro et al. have studied Mater-Bi/wood flour composites biodegradation in active sewage sludge, finding that the composites undergo biodegradation with higher weight loss rates than the neat Mater-Bi. This effect was attributed primarily to the morphology achieved and to the capability of wood fibers to act as support for the bacterial growth.

Another interesting thermoplastic polymer coming from renewable sources is polylactic acid (PLA). In general, PLA shows good mechanical properties, is biocompatible and is rather easy to produce. Literature reports some studies on PLA and natural organic fillers. Nishino et al. prepared PLA/kenaf fibers composites with good mechanical properties, thanks to the orientation imparted to the fibers. Lee and Wang have studied PLA/bamboo fibers composites and the effect of a lysine-based coupling agent, obtaining an improvement of the mechanical properties and an increased thermal stability, even though this may worsen the biodegradability. Further studies exist on PLA/flax composites and the main problem encountered regards matrix-fibers adhesion.

Huda et al. investigated the properties of PLA/recycled cellulose composites prepared by extrusion and injection molding, finding that the filler (up to 30 wt.%) significantly improved the rigidity without affecting the crystallinity degree or thermal stability.

Plackett et al. prepared PLA/jute composites, by a film stacking technique, finding significant improvements of the tensile properties, although brittle fracture was observed, affecting also the impact strength.

Unfortunately, some limitations have still to be overcome in order to support the development and use of fully biodegradable composites. First comes the still high price of biodegradable polymers in comparison to traditional commodity polymers, which of course discourages their use. Price is slowly decreasing upon increasing of their utilization, however the latter need to increase more and more to get to very competitive prices; this is one of the major challenges that biodegradable polymers are facing.

Another problem regards, as predictable, filler dispersion and its interface adhesion with the polymer matrix, which are of fundamental importance for the overall properties of the product. Depending on the biodegradable polymer used, the results maybe more satisfactory than in the case of polyolefins, etc. because of the presence of polar groups along the macromolecules but, sometimes, it is necessary to improve the properties by treatments such as alkali treatment, acetylation, and silane or maleic anhydride treatments. However, from the point of view of industrial applications, chemical modification of fibers is usually neither convenient nor cheap, therefore being mostly a niche solution.

Finally, it must be pointed out that thermoplastic polymers have been investigated also for use with nano-sized fillers in replacement for traditional micro-sized ones, thus leading to the "green nanocomposites" field. An extensive review of such subfield would be out of the scope of the present paper. However, excellent reviews on it currently exist. Samir et al. reported on the processing and behavior of nanocomposites obtained by thermoplastic polymers and polysaccharide microcrystals, finding that the use of high aspect ratio cellulose whiskers can lead to significant improvements of the mechanical properties. Polysaccharide nanocrystals may be obtained from several sources such as, for instance, chitin and starch by acid hydrolysis.

Hubbe et al. focused on possible industrial application of cellulose nanocrystals, pointing out that retention of properties over time should be guaranteed and the use of water-miscible polymer matrices such as latex, starch products, polyvinyl alcohol should be preferred, in order to make cellulose preparation and compatibilization with the matrix easier. Eichhorn et al. reported (similarly as in the previous references cited) the possible methods of cellulose nanofillers recovery, then focusing on the use of cellulose nano-whiskers for the manufacturing of shape memory nanocomposites, as well as on the interfacial phenomena occurring in polymer/nanocellulose filler composites.

[*Selected from* La Mantia F P, Morreale M. Green composites: A brief review[J]. Composites: A, 42 (2011), 579-588.]

New words and expressions

talc *n.* 滑石

kaolin *n.* 高岭土

mica *n.* 云母

wollastonite *n.* 硅酸钙岩矿

aramidic *n.* 芳族聚酰胺纤维

Notes

(1) oil palm empty fruit bunch fiber　油棕榈空果束纤维

(2) corn plants extracted fiber　玉米植株提取纤维

(3) (petroleum-based) polymer 基于石油合成的聚合物

(4) hydrolytic chain-scission reaction 水解断链反应

(5) lysine based coupling 赖氨酸基耦合

(6) a niche solution 利基型解决方案

Exercises

1. Question for discussion

(1) It is widely known that the use of a polymer and one (or more) solid fillers allows obtaining several advantages and, in particular, a combination of the main properties of the two (or more) solid phases. However, this leads also to one of the main limitations of polymer composites. What is the limitation?

(2) As a general rule, how many types of materials which the biodegradable polymers can be mainly classified into (according to their origin)?

(3) The possible industrial application of cellulose nanocrystals was focused on. The retention of properties over time should be guaranteed the use of water-miscible polymer matrices such as latex, starch products, polyvinyl alcohol should be preferred. What is the purpose?

2. Translate the following into Chinese

(1) It is widely known that the use of a polymer and one (or more) solid fillers allows obtaining several advantages and, in particular, a combination of the main properties of the two (or more) solid phases.

(2) Several points support this choice; first of all, the use of natural-organic fillers in replacement for traditional mineral inorganic ones allows a considerable reduction in the use of nonbiodegradable polymers and non-renewable resources. Furthermore, these fillers are usually drawn from relatively abundant plants (often from wastes), therefore they are very cheap.

(3) Scaffaro et al. found that injection molding can significantly reduce the polymer's molecular weight in comparison to extrusion followed by compression molding but, on the other hand, it can improve both surface and internal morphology of the composite samples, thus leading to significant increases of tensile strength if compared to the neat, unfilled samples.

(4) Already explored industrial applications include window and doorframes, furniture, railroad sleepers, automotive panels and upholstery, gardening items, packaging, shelves and in general those applications which do not require very high mechanical resistance but, instead, low purchasing and maintenance costs.

(5) Several biodegradable, natural-derived polymers exist, such as polysaccharides (starch, chitin, collagen, gelatines, etc.), proteins (casein, albumin, silk, elastin, etc.), polyesters (e.g. poly(hydroxyalcanoate), poly(hydroxybutyrate), polylactic acid), lignin, lipids, natural rubber, some polyamides, polyvinyl alcohols, polyvinyl acetates, and polycaprolactone.

(6) An interesting system for automotive applications is a composite where the polymer matrix is based on soy and corn oil, which are used as raw materials for the production of the polymer (in a way which is comparable with the one by which commodity polymers come from oil) with good resistance, flexibility, lightness and durability properties.

(7) Another interesting thermoplastic polymer coming from renewable sources is polylactic acid (PLA). In general, PLA shows good mechanical properties, is biocompatible and is rather easy to produce.

(8) Price is slowly decreasing upon increasing of their utilization, however the latter need to increase more and more to get to very competitive prices; this is one of the major challenges that biodegradable polymers are facing.

(9) However, from the point of view of industrial applications, chemical modification of fibers is usually neither convenient nor cheap, therefore being mostly a niche solution.

3. Translate the following into English

(1) 此外，在生产塑料过程中会消耗大量的石油资源，它是一种不可再生资源。

(2) 木粉和纤维是最知名和最常用的天然有机填料。

(3) 木屑、木质纤维的主要缺点是填料和聚合物基质间的结合较差、抗冲击强度低、在温度高于200℃会发生热分解。

(4) 第一个使用天然纤维做聚合物填料的汽车制造商梅萨德斯-奔驰公司早在 90 年代即用含有黄麻纤维的复合物做汽车门板。

(5) 在多数状况下，它们在适当的环境（通常是潮湿环境）中通过酶化反应发生降解。

(6) 向聚合物中加入颗粒状淀粉会降低复合材料断裂处伸缩性（即应力），而淀粉含量的增加也会增强材料的弹性模量。

(7) 现有文献对复合材料在土壤中的生物降解进行了研究，发现其在土壤中埋藏后同样发生了生物降解。

(8) 可以看出，木粉显著提升了材料的刚度和抗热机械性能。

(9) 最后，需要指出的是，热塑性聚合物已经用于纳米尺寸填料以取代微米尺寸的填料，因此有了“绿色纳米复合材料”领域。

4. Scenario simulation

You're a technical engineer from basf, Infinergy is the world's first expanded thermoplastic polyurethane (E-TPU). The closed-cell, elastic particle foam combines the properties of TPU with the advantages of foams, making it as elastic as rubber but lighter.

Like its starting material TPU, Infinergy® is noted for having high elongation at break, tensile strength and abrasion resistance as well as good chemical resistance. In addition, the innovative particle foam remains highly elastic and soft over a wide temperature range.

The feature of Infinergy® that is particularly striking is its high resilience. Tests of the resilience elasticity under ISO 8307 (the ball rebound test) and under DIN 53512 (using a defined pendulum hammer) show that Infinergy® achieves a rebound of over 55%. This is therefore significantly higher than comparable particle foams such as expanded polypropylene (EPP) at 30%, ethylene vinyl acetate (EVA) at 37% or expanded polyethylene (EPE) at 50%. Infinergy® does not lose its excellent resilience even under a continuous load.

Dynamic mechanical analysis has shown that, even at extremely low temperatures of −20°C (−4°F), Infinergy® still has a low dynamic modulus, is very soft and stretchy and thus does not go stiff.

Write an e-mail to ANTA company to Explain why Infinergy® is better for sneakers.

在线习题

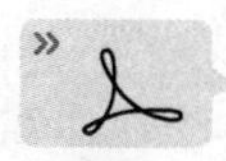
拓展阅读

Chapter 2 Matrices

Unit 1 Polymer matrix

A brief description of the various matrix materials, polymers, metals, and ceramics, is given in this chapter. We emphasize the characteristics that are relevant to composites. The reader should consult the references listed under suggested reading for greater details regarding any particular aspect of these materials.

1. Polymers

Polymer are structurally much complex than metals or ceramics. They are cheap and can be easily processed. On the other hand, polymers have lower strength and modulus and lower temperature use limits. Prolonged exposure to ultraviolet light and some solvents can cause the degradation of polymer properties. Because of predominantly covalent bounding, polymers are generally poor conduction of heat and electricity. Polymers, however, are generally more resistant to chemicals than are metals. Structurally, polymers are giant chainlike molecules (hence the name macromolecules) with covalently bonded carbon atoms forming the backbone of the chain. The process of forming large molecules from small ones is called polymerization; that is polymerization is the process of joining many monomers, the basic building blocks, together to form polymers. There are two important classes of polymerization.

① **Condensation polymerization** In this process a stepwise reaction of molecules occurs and in each step a molecule of a simple compound, generally water, forms as a by-product.

② **Addition polymerization** In this process monomers join to form a polymer without producing any by-product. Addition polymerization is generally carried out in the presence of catalysts. The linear addition of ethylene molecules (CH_2) results in polyethylene (a chain of ethylene molecules, Fig. 2.1), with the final mass of polymer being the sum of monomer masses:

Based on their behavior, there are two major classes of polymers, produced by condensation or addition polymerization, i.e., thermosetting and thermoplastic polymers. Thermosets undergo a curing reaction that involves crosslinking of polymeric chains. They harden on curing, hence the term thermoset. The curing reaction can be initiated by appropriate chemical agents or by application of heat and pressure, or by exposing the monomer to an electron beam. Thermoplas-

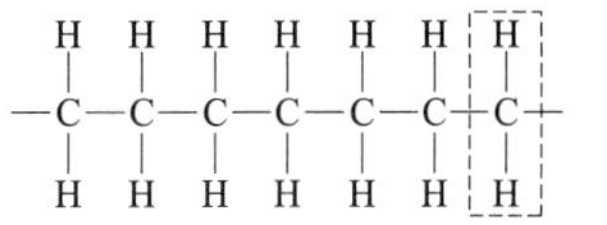

```
 H   H   H   H   H   H ┆ H ┆
 |   |   |   |   |   | ┆ | ┆
—C—C—C—C—C—C—┼C—┼
 |   |   |   |   |   | ┆ | ┆
 H   H   H   H   H   H ┆ H ┆
```

Fig. 2.1 Polyethylene structure

tics are polymers that flow under the application of heat and pressure, i.e; they soften or become plastic on heating, cooling to room temperature hardens thermoplastics. Their different behavior, however, stems from their molecular structure and shape, molecular size or mass, and the amount and type of bond (covalent or van der Waals). We first describe the basic molecular structure in terms of the configuration of chain molecules. Fig. 2.2 shows the different chain configuration types.

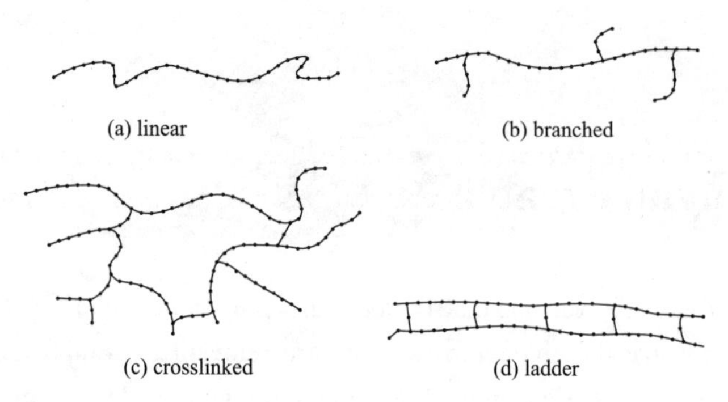

Fig. 2.2 Different molecular chain configurations

(1) Linear polymers As the name suggests, this type of polymer consists of a long chain of atoms with attached side groups. Examples include polyethylene, polyvinyl chloride, and polymethyl methacrylate. Fig. 2.2(a) shows the configuration of linear polymer; note the coiling and bending of chains.

(2) Branched polymers Polymer branching can occur with linear, crosslinked, or any other type of polymer; see Fig. 2.2(b).

(3) Crosslinked polymers In this case, molecules of one chain are bonded with those of another; see Fig. 2.2(c). Crosslinked of molecular chains results in a three-dimensional network. Crosslinking of makes sliding of molecules past one another difficult, thus such polymers are strong and rigid.

(4) Ladder polymers If we have two linear polymers linked in a regular manner [Fig. 2.2(d)] we get a ladder polymer. Not unexpectedly, ladder polymers are more rigid than linear polymers.

2. Glass transition temperature

Pure crystalline materials have well-defined melting temperatures. The melting point is the temperature at which crystalline order is completely destroyed on heating. Polymers, however, show a range of temperatures over which crystallinity vanishes. Fig. 2.3 shows specific volume (volume/unit mass) versus temperature curves for amorphous and semicrystalline polymers. When a polymer liquid is cooled, it contracts. The contraction occurs because of a decrease in the thermal vibration of molecules and a reduction in the

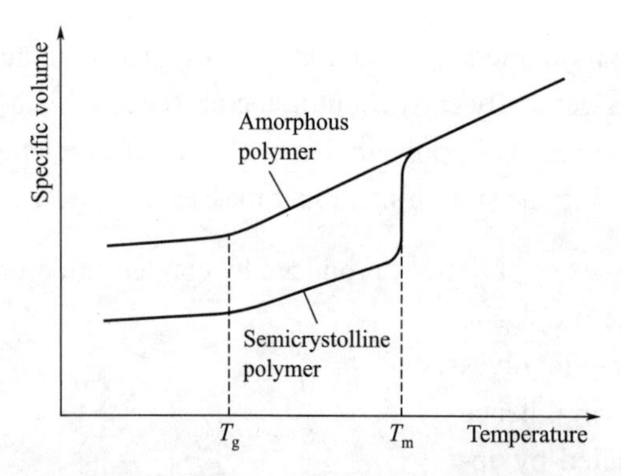

Fig. 2.3 Specific volume versus temperature for amorphous and semicrystalline polymers

free space; that is, the molecules occupy the space less loosely. In the case of amorphous polymers, this contraction continues below T_m, the melting point of crystalline polymer, to T_g the glass transition temperature where the supercooled liquid polymer becomes extremely rigid owing to extremely high viscosity. Unlike the melting point, T_m, where there occurs a transformation from the liquid phase to a crystalline, the structure of a glassy or amorphous material below T_g is essentially that of a liquid, albeit a very viscous one. Such a phenomenon is commonly observer in silica-based inorganic glasses. In the case of amorphous polymers, we are dealing with a glassy structure made of organic molecules. The glass transition temperature T_g, although it does not represent a thermodynamic phase transformation, is in many ways akin to the melting point for the crystalline solids. Many physical properties (e.g., viscosity, heat capacity, elastic modulus, and expansion coefficient) change abruptly at T_g. Polystyrene, for example, has a T_g of about 100 ℃ and is therefore rigid at room temperature. Rubber, on the other hand, has a T_g of about −75 ℃ and therefore is flexible at room temperature. T_g is a function of the chemical structure of the polymer. For example, if a polymer has a rigid backbone structure and/or bulky branch groups, then T_g will be quite high.

Although both amorphous polymers and inorganic silica-based glasses have a glass transition temperature T_g, generally, the T_g of inorganic glasses is several hundred degrees Celsius higher than that of polymers. The reason for this is the different types of bonding and the amount of crosslinking in the polymers and glasses. Inorganic glasses have mixed covalent and ionic bonding and are highly crosslinked. This gives them a higher thermal stability than polymers which have covalent and van der Waals bonding and a lesser amount of crosslinking than found in inorganic glasses.

New words and expressions

modulus *n.* 模量
ultraviolet *n.* 紫外线; *adj.* 紫外线的; *v.* 紫外线辐射
predominantly *adv.* 占主导地，占优势地，主要地
macromolecule *n.* 大分子
polymerization *n.* 聚合反应，聚合作用
condensation *n.* 冷凝，凝结，缩聚，缩合作用
by-product *n.* 副产物
ethylene *n.* 乙烯
catalyst *n.* 催化剂
polyethylene *n.* 聚乙烯
thermosetting *adj.* 热固性的，热凝性的，热成形的
covalent *adj.* 共有原子价的，共价的
polyvinyl chloride 聚氯乙烯
polymethyl methacrylate 聚甲基丙烯酸甲酯
configuration *n.* 组态，构造，结构，配置，外形
crystallinity *n.* 结晶度，结晶性
supercool *v.* 使过度冷却
amorphous *adj.* 无定形的
akin *adj.* 性质相同的，类似的
polystyrene *n.* 聚苯乙烯
flexible *adj.* 韧性的，挠性的
silica-based *adj.* 硅基的，石英基的
Celsius *n.* 摄氏度; *adj.* 摄氏度的

Notes

(1) The process of forming large molecules from small ones is called polymerization; that is polymerization is the process of joining many monomers, the basic building blocks, together to form polymers.
由小分子形成大分子的过程称为聚合；或者说聚合是把很多基本组成——单体构成聚合物的过程。

(2) Condensation polymerization: In this process a stepwise reaction of molecules occurs and in each step a molecule of a simple compound, generally water, forms as a by-product.
缩聚反应：在这个过程中分子发生逐步反应，每步反应会有一个简单的化合物作为副产物产生，一般是水。

(3) Thermoplastics are polymers that flow under the application of heat and pressure, i.e., they soften or become plastic on heating, cooling to room temperature hardens thermoplastics.
热塑性塑料是在受热和受压下具有流动性的聚合物，亦即，它们受热会软化并具有塑性，冷却至室温会变硬。

(4) The contraction occurs because of a decrease in the thermal vibration of molecules and a reduction in the free space; that is, the molecules occupy the space less loosely.
收缩的出现是由于分子热振动的下降和自由空间的减少，也就是说，分子占据的空间变少了。

(5) The glass transition temperature T_g, although it does not represent a thermodynamic phase transformation, is in many ways akin to the melting point for the crystalline solids.
玻璃化转变温度 T_g，尽管它并不发生热力学相变，但在很多方面类似于晶体的熔点。

Exercises

1. Question for discussion

(1) What is a composite?
(2) What is a polymer?
(3) Where are polymers widely used?
(4) Give the difference of two important classes of polymerization.
(5) What is the T_g value of the polystyrene?

2. Translate the following into Chinese

(1) Polymer are structurally much complex than metals or ceramics.
(2) Structurally, polymers are giant chainlike molecules (hence the name macromolecules) with covalently bonded carbon atoms forming the backbone of the chain.
(3) Because of predominantly covalent bounding, polymers are generally poor conduction of heat and electricity.
(4) Based on their behavior, there are two major classes of polymers, produced by either condensation or addition polymerization, i.e., thermosetting and thermoplastic polymers.
(5) Their different behavior, however, stems from their molecular structure and shape, molecular size or mass, and the amount and type of bond (covalent or van der Waals).
(6) In the case of amorphous polymers, this contraction continues below T_m, the melting point of crystalline polymer, to T_g the glass transition temperature where the supercooled liquid polymer becomes extremely rigid owing to extremely high viscosity.
(7) Crosslinking of makes sliding of molecules past one another difficult, thus such polymers are strong and rigid.
(8) Although both amorphous polymers and inorganic silica-based glasses have a glass transition

temperature T_g, generally, the T_g of inorganic glasses is several hundred degrees Celsius higher than that of polymers.

inorganic glass
silica-based glasses
amorphous polymers
supercooled liquid polymer
thermoplastic polymer
thermosetting polymer
chainlike molecule
thermal vibration

3. Translate the following into English

聚合　　大分子　　无定形的
缩聚反应　　副产物　　玻璃化转变温度
热力学相变　　催化剂

(1) 聚合物结构比金属和陶瓷要复杂。
(2) 加聚反应一般需要催化剂。
(3) 许多物理性质会在物质的玻璃化温度发生突变。
(4) 这种现象常常在硅基无机玻璃被观察到。
(5) 热固性材料的固化反应涉及到聚合链的交联。
(6) 发生收缩是由于分子热振动的减少和自由空间的减少;也就是说，分子占据的空间不那么松散。
(7) 这使得它们比具有共价键和范德华键的聚合物具有更高的热稳定性，并且比无机玻璃中的交联量更少。

4. Scenario simulation

Suppose you're an engineer in a composite material factory, you are responsible for introducing some basic concepts about composite materials to the workers. Please prepare an outline for your powerpoint.

在线习题

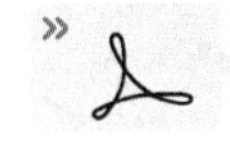
拓展阅读

Unit 2 Ceramic matrix

Ceramic materials are very hard and brittle. Generally, they consist of one or more metals combined with a nonmetal such as oxygen, carbon, or nitrogen. They have strong covalent and ionic bonds and very few slip systems available compared to metals. Thus, characteristically, ceramics have low failure strains and low toughness or fracture energies. In addition to being brittle, they lack uniformity in properties, have low thermal and mechanical shock resistance, and have low tensile strength. On the other hand, ceramic materials have very high elastic moduli, low densities, and can withstand very high temperatures. The last item is very important and is the real driving force behind the effort to produce tough ceramics. Consider the fact that up to 800 ℃ and can go up to 1100 ℃ with oxidation-resistant coatings. Beyond this temperature, one must use ceramic materials.

By far, the major disadvantage of ceramics is their extreme brittleness. Even the minutest of surface flaws (scratches or nicks) or internal flaws (inclusions, pores, or micro cracks) can have disastrous results. One important approach to toughen ceramics involves fiber reinforcement of brittle ceramics. We shall describe the ceramic matrix composites in another chapter. Here we make a brief survey of ceramic materials, emphasizing the ones that are commonly used as matrices.

1. Bonding and Structure

Ceramic materials, with the exception of glasses, are crystalline, as are metals. Unlike metals, however, ceramic materials have mostly ionic bonding and some covalent bonding. Ionic bonding involves electron transfer between atomic species constituting the ceramic compound; that is, one atom gives up an electron(s) while another accepts an electron(s). Electrical neutrality is maintained; that is, positively charged ions (*cations*) balance the negatively charged ions (*anions*). Generally, ceramic compounds are stoi-chiometric; that is, there exists a fixed ratio of cations to anions. Examples are alumina (Al_2O_3), beryllia (BeO), spinels ($MgAl_2O_4$), silicon carbide (SiC), and silicon nitride (Si_3N_4). It is not uncommon, however, to have nonstoichiometric ceramic compounds, for example, $Fe_{0.96}O$. The oxygen ion (*anion*) is generally very large compared to the metal ion (*cation*). Thus, cations occupy interstitial positions in a crystalline array of anions.

Crystalline ceramics generally exhibit close- packed cubic and hexagonal close-packed structures. The simple cubic structure is also called the cesium chloride structure. It is, however, not very common. CsCl, CsBr, and CsI show this structure. The two species form an interpenetrating cubic array, with anions occupying the cube corner positions while cations go to the interstitial sites. Cubic close packed is really a variation of the fcc structure described in section above. Oxygen ions (anions) make the proper fcc structure with metal ions (cations) in the interstices. Many ceramic materials show this structure, also called the NaCl or rock salk-type structure. Examples include MgO, CaO, FeO, NiO, MnO, and BaO. There are other variations of fcc closed-packed structures, for example, zinc blende types (ZnS) and fluorite types (CaF). The hexagonal close-packed structure is

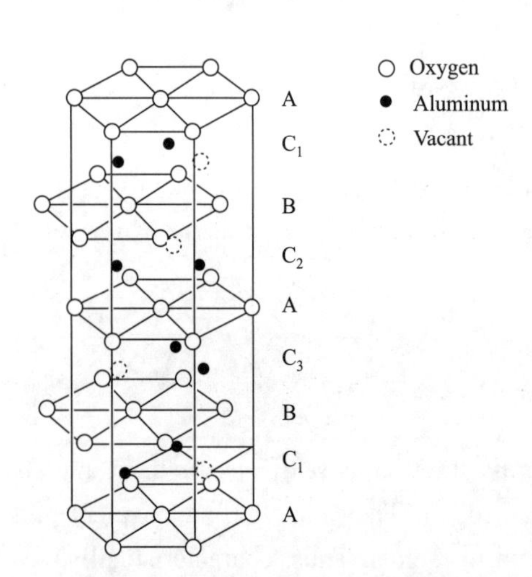

Fig. 2.4 Hexagonal closed-packed structure of α-alumina.

also cry crystallizes in the hcp form. Other examples are nickel arsenide (NiAs) and corundum (Al_2O_3). Fig. 2.4 shows the hcp crystal structure of α-Al_2O_3. A and B layers consist of oxygen atoms while C_1, C_2, and C_3 layers contain aluminum atoms. The C layers are only two-thirds full.

Glass-ceramic materials form yet another important category of ceramics. They form a sort of composite material because they consist of 95%-98% by volume of crystalline phase and the rest glassy phase. The crystalline phase is very fine (grain size less than 1 μm in diameter). Such a fine grain size is obtained by adding nucleating agents (commonly TiO_2 and ZrO_2) during the melting operation, followed by controlled crystallization. Important examples of glass-ceramic systems include:

① **Li_2O-Al_2O_3-SiO_2** This has a very low thermal expansion and is therefore very resistant to thermal shock, corningware is a well-known trade name of this class of glass-ceramic;

② **MgO-Al_2O_3-SiO_2** This has high electrical resistance coupled with high mechanical strength.

Ceramic materials can also form solid solutions. Unlike metals, however, interstitial solid solutions are less likely in ceramics because the normal interstitial sites are already filled. Introduction of solute ions disrupts the charge neutrality. Vacancies accommodate the unbalanced charge. For example, FeO has a NaCl-type structure with an equal number of Fe^{2+} and O^{2-} ions. If, however, two Fe^{3+} ions were to replace three Fe^{2+} ions we would have a vacancy where an iron ion would form.

Glasses, the traditional silicate ceramic materials, are inorganic solid-like materials that do not crystallize when cooled from the liquid state. Their structure is not crystalline but that of a supercooled liquid. In this case we have a specific volume versus temperature curve similar to the one for polymers and a characteristic glass transition temperature T_g. Under certain conditions, crystallization of glass can occur with an accompanying abrupt decrease in volume at the melting point because the atoms take up ordered positions.

2. Common Ceramic Matrix Materials

Silicon carbide has excellent high-temperature resistance. The major problem is that it is quite brittle up to very high temperature and in all environments. Silicon nitride is also an important nonoxide ceramic matrix material.

Among the oxide ceremics, alumina and mullite are quite promising. Silica-based glasses and glass-ceramics are other ceramic matrices. With glass-ceramics one can densify the matrix in a glassy state with fibers, followed by crystallization of the matrix to obtain high-temperature stability.

Ceramic matrices are used in fiber reinforced composites to achieve, in addition to high strength and stiffness, high-temperature stability and adequate fracture toughness. Tab. 2.1 summarizes some of the important characteristics of common ceramic matrix materials (Phillips. 1983).

Tab. 2.1 Properties of some ceramic matrix materials

Material	Young's modulus /GPa	Tensile strength /MPa	Coefficient of thermal expansion $/\times10^{-6}$/K	Density g/cm^3
Borosilcate glass	60	100	3.5	2.3
Soda glass	60	100	8.9	2.5
Lithium aluminosilicate glass-ceramic	100	100~150	1.5	2.0
Magnesium aluminosilicate glass-ceramic	120	110~170	2.5~5.5	2.6~2.8
Mullite	143	83	5.3	
MgO	210~300	97~130	13.8	3.6
Si_3N_4	310	410	2.25~2.87	3.2
Al_2O_3	360~400	250~300	8.5	3.9~4.0
SiC	400~440	310	4.8	3.2

Source: Adapted with permission from Phillips (1983).

New words and expressions

ceramic *n.* 陶瓷；*a.* 陶瓷的，陶器的
slip systems 滑移系
failure strain 失效应变
toughness *n.* 韧性，强硬
uniformity *n.* 均质性
tensile strength 拉伸强度
disadvantage *n.* 坏处，弊病；*v.* 危害
brittleness *n.* 脆性
minutest *adj.* 微小的
flaw *n.* 缺陷，裂纹
inclusion *n.* 掺杂物，杂质；*v.* 夹杂
disastrous *adj.* 灾难性的
neutrality *n.* 中性
cation *n.* 阳离子，正离子
anion *n.* 阳离子，负离子
stoichiometric *adj.* 化学计量的
interstice *n.* 间隙，空隙，小缝
close-packed *v.* 紧密堆积；*adj.* 紧密堆积的
hexagonal *adj.* 六角形的，六边形的
solid solution 固溶体，固体溶液
silicon carbide *n.* 金刚砂，碳化硅
mullite *n.* 莫来石
densify *v.* 压实，增浓，使增加密度

Notes

(1) They have strong covalent and ionic bonds and very few slip systems available compared to metals.
相比于金属，它们有强烈的共价键和离子键，但有用的滑移系很少。

(2) On the other hand, ceramic materials have very high elastic moduli, low densities, and can withstand very high temperatures.
另一方面，陶瓷材料具有非常高的弹性模量、低密度，并能承受非常高的温度。

(3) Ionic bonding involves electron transfer between atomic species constituting the ceramic compound; that is, one atom gives up an electron(s) while another accepts an electron(s).
离子键在构成陶瓷复合物的不同原子种类之间存在电子转移，即一类提供电子，而另一类接受电子。

(4) Such a fine grain size is obtained by adding nucleating agents (commonly TiO_2 and ZrO_2) during the melting operation, followed by controlled crystallization.

这种细粒度是通过在熔融操作过程中添加成核剂（通常是 TiO_2 和 ZrO_2）而形成的，紧接着是可控结晶。

(5) Ceramic matrices are used in fiber reinforced composites to achieve, in addition to high strength and stiffness, high-temperature stability and adequate fracture toughness.
除高强度和高刚度外，陶瓷基体用于纤维增强复合材料中，还可起到提高高温稳定性和断裂韧性的作用。

Exercises

1. Question for discussion

(1) What is a ceramic matrix?
(2) Give some examples of common ceramic matrix composites.

2. Translate the following into Chinese

(1) Even the minutest of surface flaws (scratches or nicks) or internal flaws (inclusions, pores, or micro cracks) can have disastrous results.
(2) On the other hand, ceramic materials have very high elastic moduli, low densities, and can withstand very high temperatures.
(3) One important approach to toughen ceramics involves fiber reinforcement of brittle ceramics.
(4) They form a sort of composite material because they consist of 95%-98% by volume of crystalline phase and the rest glassy phase.
(5) Under certain conditions, crystallization of glass can occur with an accompanying abrupt decrease in volume at the melting point because the atoms take up ordered positions.
(6) Crystalline ceramics generally exhibit close-packed cubic and hexagonal close-packed structures.
(7) Glasses, the traditional silicate ceramic materials, are inorganic solid-like materials that do not crystallize when cooled from the liquid state.
(8) With glass-ceramics one can densify the matrix in a glassy state with fibers, followed by crystallization of the matrix to obtain high-temperature stability.

ionic bonding	crystalline array	glass transition temperature
covalent bonding	cubic structure	silicon carbide
electrical neutrality	crystalline phase	

3. Translate the following into English

电子转移	热膨胀系数	耐高温性
晶体阵列	机械强度	断裂韧性
玻璃相	熔点	

(1) 陶瓷材料坚硬而脆。
(2) 它们由一种或多种金属与非金属（氧、碳、氮）结合而构成。
(3) 这种简单的晶胞结构叫做氯化铯结构。
(4) 碳化硅有极好的耐高温性能。
(5) 在一定的条件下，由于原子处于有序的位置，玻璃在熔点处的体积会随之急剧减小，从而发生结晶。

(6) 氮化硅也是一种重要的非氧化物陶瓷基体材料。
(7) 陶瓷基体用于纤维增强复合材料，除了具有高强度和刚度、高温稳定性和足够的断裂韧性外。

Unit 3 Metal matrix

A wide range of metals and their alloys may be used as matrix materials. In this chapter, we review some of the basic concepts and fundamentals of bonding and structure of common metals. Following this, we provide a summary of the characteristics of some of the most common metals that are used as matrix materials in metal matrix composites.

1. Bonding and Crystalline Structure in Metals

Metals are characterized by metallic bonding, i.e., valence electrons are not bound to a particular ion in the solid. A "sea of electrons" surrounds positively charged atomic nuclei. A major result of this electron cloud surrounding the atomic nuclei is that the electronic bonding in metals is nondirectional. This non-directionally of bonding is very important since it contributes to isotropy in many properties. When cooled down from their molten state, most metals assume a crystalline structure below their melting point. Some metals and alloys do not undergo crystallization and assume an amorphous structure, but only at very high cooling rates (> 10^6 K/s). The basic difference between a crystalline and non-crystalline structure is the degree of ordering. A fully crystalline state has a high degree of order. Metal ions are quite small (diameter about 0.25 nm), so in a crystalline structure these ions are packed in a very regular and close packed manner. Because of non-directional bonding, we can model the arrangement of atoms in the form of hard spheres. There are two arrangements of packing of hard, identical spheres that result in close packed structures: Face centered cubic (FCC) and hexagonal close packed (HCP). There is a third arrangement that is observed in a number of metals, namely, body centered cubic (BCC). The BCC structure is more open than the FCC or HCP structure. Fig. 2.5 shows the atomic arrangement in these three cases. Both HCP and FCC structures result in close-packed structures. The difference between the two results from the way the close packed planes are stacked. Fig. 2.6(a) shows the stacking arrangement of the first layer (layer **A**) in a close packed structure. For the second layer of close packed atoms, one has the choice of putting the atoms at sites **B** or **C**, which are equivalent sites. The difference between FCC and HCP structures comes in the placement of the third layer of atoms. Assuming the B configuration for the second layer of atoms, one can have the third layer in A or C configuration. It turns out that the **ABABAB**... (or **ACACAC**...) results in HCP while the **ABCABCABC**... stacking sequence results in the FCC structure. Stacking faults or planar defects may be formed, where there is an interruption of **ABCABCABC** stacking (FCC) sequence such that we have an HCP region in an FCC structure.

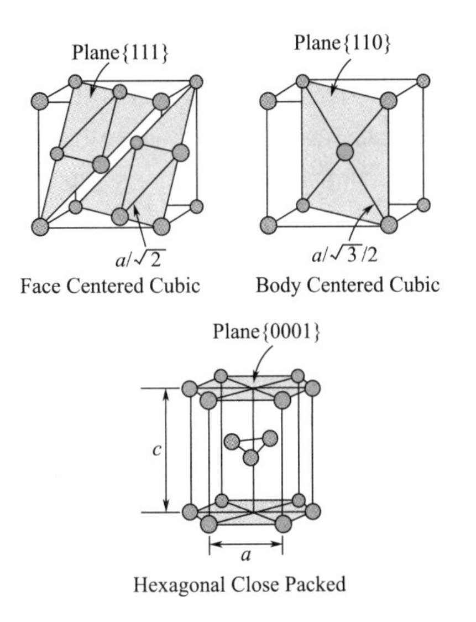

Fig. 2.5 Three common crystal structures in metals, face centered cubic (FCC), body centered cubic (BCC), and hexagonal close packed (HCP)

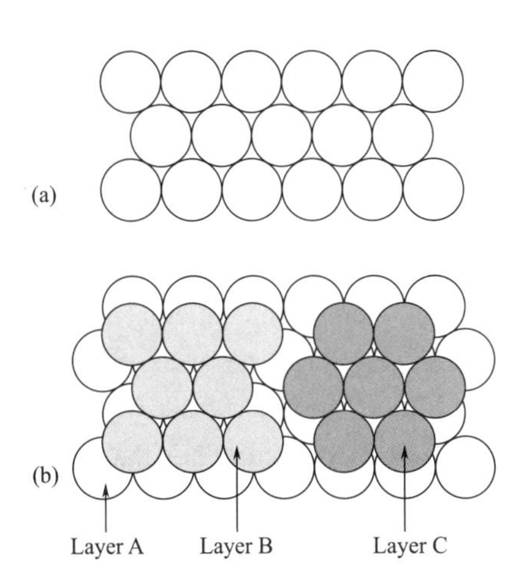

Fig. 2.6 (a) **A** layer of close packed atoms (b) Positions of layers **B** and **C** on top of layer **A**. Close packed layers in FCC has **ABCABC** stacking, while HCP has **ABAB**... stacking sequence.

2. Crystalline defects in metals

Most metals are crystalline in their solid state. Ideally, a metal crystal has atoms arranged in a very orderly fashion in a three dimensional lattice. Such perfect crystals are rarely seen in practice. Generally, various kinds of imperfections are present. It turns out that many of the important and interesting properties of crystalline materials are due to the presence of these imperfections or defects in the crystalline lattice. These defects, called lattice defects, can be zero-, uni-, bi-, or tri-dimensional. The zero-dimensional defects are an extension of the order of the lattice constant (0.2-0.4 nm) in all three spatial directions. Common examples are vacancies and interstitial atoms, etc. Unidimensional or linear imperfections have atomic extensions in one spatial direction only. The most important example of a linear defect is a dislocation. Bi-dimensional defects have atomic extension in two spatial directions, for example, grain boundaries, twin boundaries, interfaces between phases, etc. Three-dimensional defects extend into three spatial directions, for example, second phase particles, pores, etc. These defects may result under less than ideal processing conditions. Besides geometric differences, there is a very important physical difference between zero-dimensional defects and other higher-dimensional defects in crystals. Considering the free energy involved, only point defects (i.e., zero-dimensional) can be present in a state of thermal equilibrium in a crystal, while dislocations, grain boundaries, etc., represent energy much higher than energy represented by thermal fluctuation. ***A*** dislocation can be described in terms of two vectors.

① **Dislocation line vector, *t*** which gives the direction of the dislocation line at any point. The vector t is parallel to the dislocation. In the case of a dislocation loop, the vector *t* runs along the loop length and will have opposite senses along the two opposite sides of the loop.

② **Burgers vector, *b*** This indicates the magnitude and direction of the displacement of the part above the slip plane of a crystal with respect to the part below the slip plane. The Burgers vector ***b*** of a dislocation loop is constant, while the dislocation line vector ***t*** can change direction continuously. The Burgers vector is always an integral atomic spacing because the lattice must maintain atomic registry through the slipped and unslipped regions. There are two special types of dislocations: (a) Edge dislocation: the dislocation line vector and the Burgers vector are orthogonal, i.e., at ninety degrees; (b) Screw dislocation: the dislocation line vector and the Burgers vector are parallel.

Dislocations control many important characteristics of materials. In particular, dislocations occupy a position of fundamental importance in the mechanical behavior of crystalline materials. Their presence reduces the force necessary to cause displacement of atoms. In this sense, dislocations act like a lever. They allow a given quantity of work to be done by a small force moving through a large distance rather than a large force moving through a small distance. Thus, plastic flow or plastic deformation in crystalline solids is accomplished by means of movement of dislocations. The plastic strain depends on the displacement caused by each dislocation, the density of *mobile* dislocations, and the average distance moved by a dislocation. When these linear defects move under the action of a shear stress, they result in slip or glide between crystal planes, which in turn results in permanent (plastic) deformation. Besides plastic deformation, there are many other physical and chemical properties that are affected by the presence of dislocations. For example, dislocations can serve as easy paths for atomic diffusion (which can affect creep behavior), precipitation reactions, and order processes. They can also be very efficient sites for nucleation of solid-state phase transformations, affect thermal and electrical conductivity, especially at very low temperatures, and affect the current carrying capacity in superconductors. Planar or two-dimensional defects include external surfaces, grain boundaries, stacking faults, etc. The surface is treated as a planar imperfection because the surface atoms are not bonded to the maximum number of neighbors. Thus, the surface atoms are in a higher energy state than the atoms in the interior. Small angle grain boundaries consist of slight orientation mismatch (few degrees). Simple dislocation arrays form such boundaries; a wall of aligned edge dislocations forms a tilt boundary while screw dislocations form twist boundaries. When the degree of misorientation between grains is too large to be accommodated by dislocations, high angle grain boundaries are formed. These high angle grain boundaries have a higher energy at the grain boundaries than small angle grain boundaries. A twin boundary is a special type of grain boundary across which there is mirror symmetry of the lattice. The region of material between these boundaries is termed a twin. Twin boundaries can result from mechanical shear (mechanical twins) or annealing following deformation (annealing twins). Annealing twins are commonly observed in FCC crystals while mechanical twins are seen in BCC and HCP metals.

New words and expressions

valence *n.* 化合价
nondirectional *adj.* 无方向的，非定向的，适合各方向的
contribute *v.* 捐献，贡献，添加
equivalent *adj.* 相同的，同等的；*n.* 同等物
imperfection *n.* 缺陷，不足之处，不完整性
lattice *n.* 格子金属架，晶格，点阵
unidimensional *adj.* 线性的，一维的，一度空间的
grain boundary 晶界
dislocation *n.* 位错，混乱
vector *n.* 矢量
loop *n.* 环，环路，活套
screw *n.* 螺杆，螺钉
deformation *n.* 变形，畸变
shear stress 剪应力
tilt *n.* 倾斜，倾侧
annealing *n.* 退火，韧化

Notes

(1) A major result of this electron cloud surrounding the atomic nuclei is that the electronic bonding in metals is nondirectional.
电子云围绕原子核运动的一个主要结果是金属中的电子键是非定向的。

(2) There are two arrangements of packing of hard, identical spheres that result in close packed structures: Face centered cubic (FCC) and hexagonal close packed (HCP).
等体积硬球有两种堆积方式从而导致有两种密堆积结构：面心立方堆积（FCC）和六方密堆积（HCP）。

(3) Stacking faults or planar defects may be formed, where there is an interruption of ABCABCABC stacking (FCC) sequence such that we have an HCP region in an FCC structure.
ABCABCABC 堆叠序列的中断能导致堆积层错或面缺陷的形成，从而在一个 FCC 结构中就产生了 HCP 结构区域。

(4) The Burgers vector is always an integral atomic spacing because the lattice must maintain atomic registry through the slipped and unslipped regions.
伯格斯矢量始终是原子间距的整数倍，因为晶格在滑移和未滑移区域必须与原子的初始位置保持一致。

(5) The plastic strain depends on the displacement caused by each dislocation, the density of mobile dislocations, and the average distance moved by a dislocation.
塑性应变取决于每一个位错引起的位移、移动位错密度以及位错移动的平均距离。

(6) They can also be very efficient sites for nucleation of solid-state phase transformations, affect thermal and electrical conductivity, especially at very low temperatures, and affect the current carrying capacity in superconductors.
它们对于固态相变的成核也是非常有效的，影响导热性和导电性，特别在非常低的温度下，并且影响超导体的电流承载能力。

Exercises

1. Question for discussion

(1) What are metals characterized by?
(2) What are three kinds of crystalline structures in metals?

2. Translate the following into Chinese

(1) Metals are characterized by metallic bonding, i.e., valence electrons are not bound to a particular ion in the solid. A "sea of electrons" surrounds positively charged atomic nuclei.
(2) This non-directionally of bonding is very important since it contributes to isotropy in many properties.
(3) Metal ions are quite small (diameter~0.25 nm), so in a crystalline structure these ions are packed in a very regular and closepacked manner.
(4) Because of non-directional bonding, we can model the arrangement of atoms in the form of

hard spheres.

(5) It turns out that many of the important and interesting properties of crystalline materials are due to the presence of these imperfections or defects in the crystalline lattice.

(6) Thus, plastic flow or plastic deformation in crystalline solids is accomplished by means of movement of dislocations.

(7) When these linear defects move under the action of a shear stress, they result in slip or glide between crystal planes, which in turn results in permanent (plastic) deformation.

(8) Small angle grain boundaries consist of slight orientation mismatch (few degrees).

atomic nuclei　　amorphous structure　　body centered cubic　　crystalline lattice
slip plane　　plastic deformation　　shear stress　　screw dislocations

3. Translate the following into English

金属键　　金属离子　　晶格缺陷　　晶界
柏氏矢量　　蠕变行为　　导电性　　孪晶界

(1) 从围绕原子核的电子云得出的主要结论是金属中的电子键是无方向性的。

(2) 绝大多数金属从熔融态冷却下来后，在其熔点温度下呈晶体结构。

(3) HCP 和 FCC 结构导致密堆积结构。

(4) 事实上很少见到这种完美的晶体。

(5) 理想状态的金属晶体，其原子在三维晶格呈有序排布。

(6) 晶格缺陷可以是零维、一维、二维和三维。

(7) 表面的原子比内部的原子处于较高的能态。

(8) 这些高角度晶界在晶界处比小角度晶界处具有更高的能量。

在线习题

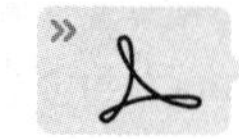
拓展阅读

Chapter 3 Reinforcements

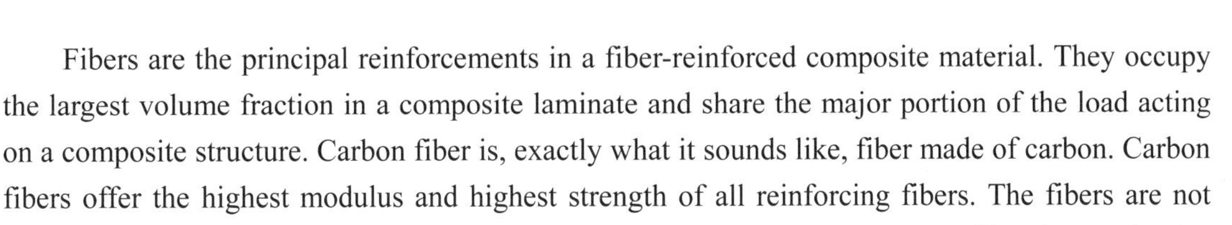

Unit 1 Introduction of carbon fibers

Fibers are the principal reinforcements in a fiber-reinforced composite material. They occupy the largest volume fraction in a composite laminate and share the major portion of the load acting on a composite structure. Carbon fiber is, exactly what it sounds like, fiber made of carbon. Carbon fibers offer the highest modulus and highest strength of all reinforcing fibers. The fibers are not susceptible to stress corrosion or stress rupture failures at room temperature unlike glass and polymeric fibers. Ongoing process development work promises significant improvements in the ratio of performance to cost which would greatly increase the uses and applications of these fibers. When bound together with plastic polymer resin by heat, pressure or in a vacuum a composite material is formed that is both strong and lightweight.

Carbon fibers have been under continuous development for the last 50 years. There has been a progression of feedstocks, starting with rayon, proceeding to polyacrylonitrile (PAN), on to isotropic and mesophase pitches, to hydrocarbon gases, to ablated graphite and finally back to carbon-containing gases. Rayon-based carbon fibers are no longer in production, and so are of historical interest only; they will not be discussed in this chapter. PAN-based fiber technologies are well developed and currently account for most commercial production of carbon fibers. Pitch-based fibers satisfy the needs of niche markets, and show promise of reducing prices to make mass markets possible. Vapour-grown fibers are entering commercial production, and carbon nanotubes are full of promise for the future.

The word "graphite" is much misused in carbon fiber literature. The word refers to a very specific structure, in which adjacent aromatic sheets overlap with one carbon atom at the center of each hexagon, as shown in Fig. 3.1. This structure appears very rarely in carbon fibers, especially in PAN-based fibers, even though they are conventionally called graphite fibers. While high-performance fibers are made up of large aromatic sheets, these are randomly oriented to each other, and are described as "turbostratic" (turbulent and stratified), as shown in Fig. 3.2. Many physical properties depend merely on the large aromatic sheets.

Because of the rich variety of carbon fibers available today, physical properties vary over a broad domain (Tab. 3.1). Fig. 3.3 shows a plot of strength versus modulus. "General purpose" fibers made from isotropic pitch have modest levels of strength and modulus. However, they are the least expensive pitch-based fiber, and are useful in enhancing modulus or conductivity in many applications. PAN-based fibers are the strongest available; however, when they are heat-treated to increase modulus, the strength decreases. Mesophase pitch fibers may be heat treated to very high modulus values, approaching the in-plane modulus of graphite at 1 TPa.

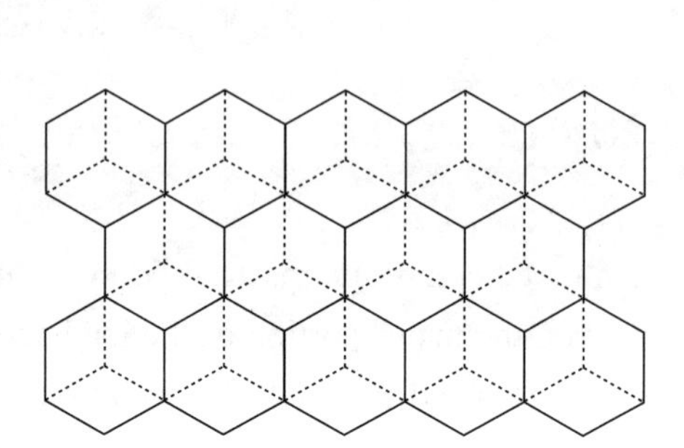

Fig. 3.1 Regular stacking of aromatic sheets in graphite

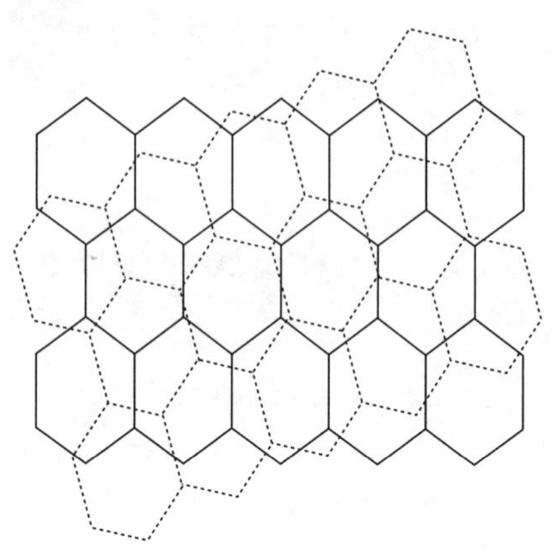

Fig. 3.2 Irregular stacking of aromatic sheets (turbostratic carbon)

Tab. 3.1 Mechanical properties of selected carbon fibers

Type	Manufacturer	Product name	Tensile strength /GPa	Young's modulus /GPa	Strain to failure /%
PAN	Toray	T300	3.53	230	1.5
		T1000	7.06	294	2.0
		M55J	3.92	540	0.7
	Hercules	IM7	5.30	276	1.8
GP-Pitch	Kureha	KCF200	0.85	42	2.1
HP-Pitch	BP-Amoco	Thornel P25	1.40	140	1.0
		Thornel P75	2.00	500	0.4
		Thornel P120	2.20	820	0.2

The Achilles' heel of mesophase pitch-based fibers in composite applications is low compressive strength; this is illustrated in Fig. 3.4. Electrical and thermal conductivity are important in many applications, and these are illustrated in Fig. 3.5 and Fig. 3.6, respectively. Mesophase pitch fibers have the highest conductivity and lowest resistivity.

Finally, there is a property of high-performance carbon fibers, both PAN and mesophase pitch-based, which sets them apart from other materials. They are not subject to creep or fatigue failure. These are important characteristics for critical applications. In a comparison of materials for tension members of tension leg platforms for deep-sea oil production, carbon fiber strand survived 2000000 stress cycles between 296 and 861 MPa. In comparison, steel pipe stressed between 21 and 220 MPa failed after 300000 cycles. Creep studies on PAN and pitch-based carbon fibers were conducted at 2300 ℃ and stresses of the order of 800 MPa. Projections of the data obtained to ambient temperatures indicate that creep deformations will be infinitesimally small.

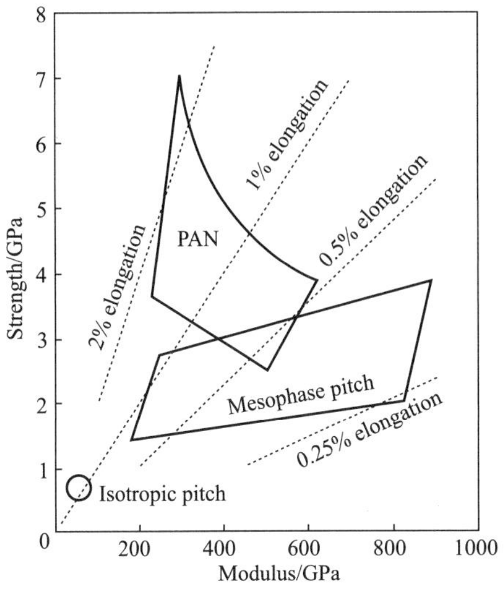

Fig. 3.3 Tensile properties of carbon fibers

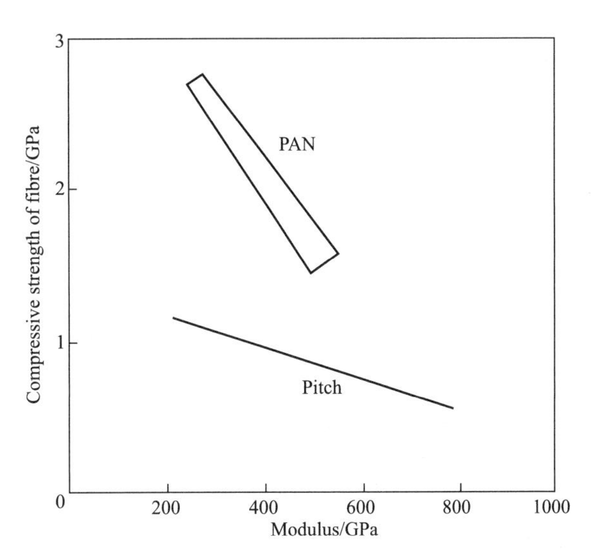

Fig. 3.4 Compressive properties of carbon fibers

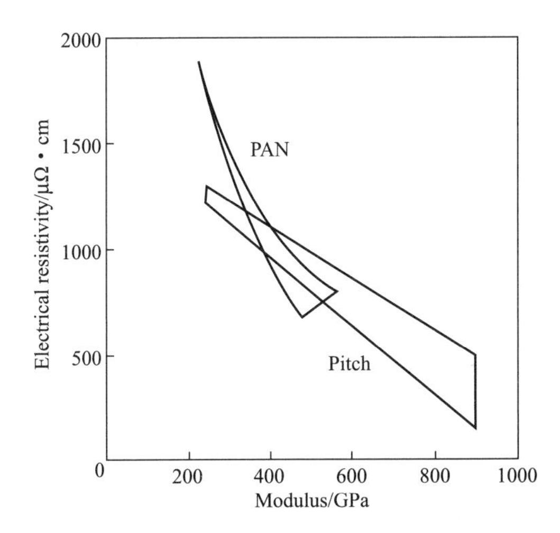

Fig. 3.5 Electrical resistivity of carbon fibers

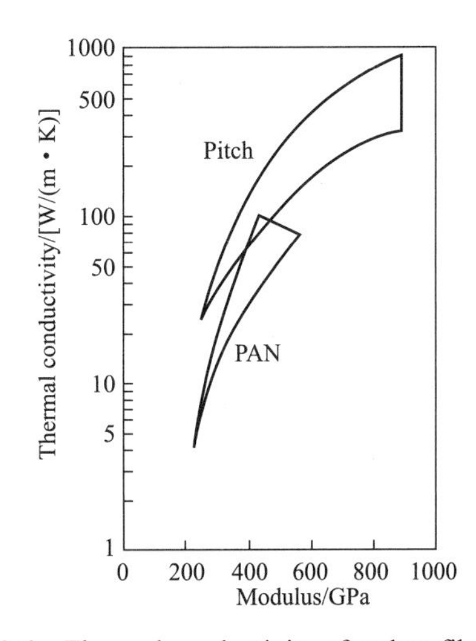

Fig. 3.6 Thermal conductivity of carbon fibers

There is a surprising degree of cross-correlation in the physical properties of carbon fibers. The mechanisms for conduction of heat and electricity are different in carbon fibers: heat is transmitted by lattice vibrations and electricity by diffusion of electrons and holes. However, there is a strong correlation between the two, as illustrated in Fig. 3.7, which allows thermal conductivity to be estimated by measurement of the electrical resistance, a much simpler measurement. Young's modulus is also correlated with electrical resistance, as shown in Fig. 3.8.

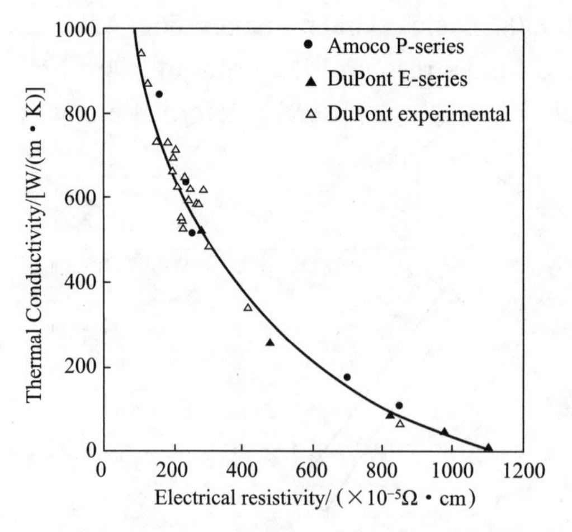

Fig. 3.7 Relationship between electrical resistivity and thermal conductivity in pitch-based carbon fibers

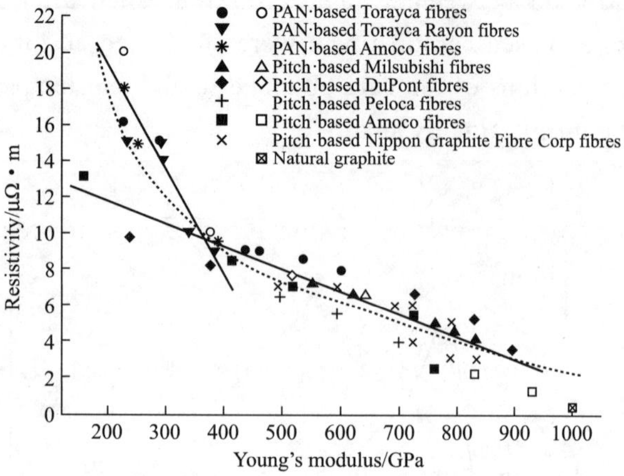

Fig. 3.8 Relationship between Young's modulus and electrical resistivity in carbon fibers

New words and expressions

feedstock *n.* 给料，原料

isotropic *adj.* 各向同性的

mesophase *n.* 中间相

ablate *v.* 烧蚀，融化

adjacent *adj.* 邻近的，毗邻的

infinitesimally *adv.* 无限小地

lattice vibration 晶格振动

Notes

(1) PAN-based fiber technologies are well developed and currently account for most commercial production of carbon fibers.
聚丙烯腈基纤维技术发展很好，目前在碳纤维产品中占主导地位。

(2) PAN-based fibers are the strongest available; however, when they are heat treated to increase modulus, the strength decreases.
聚丙烯腈基纤维是最强的可用纤维。但是，聚丙烯腈基纤维加热后模量增加，强度降低。

(3) The Achilles' heel of mesophase pitch-based fibers in composite applications is low compressive strength.
在复合材料应用中，中间相沥青基纤维最大的弱点是低抗压强度。备注：Achilles' heel 本意为“阿喀琉斯之踵”，在此翻译为“最大的弱点”。

Exercises

1. Question for discussion

(1) How many kinds of carbon fibers?

(2) Why the word 'graphite' is much misused in carbon fiber literature?

(3) What are important characteristics of carbon fibers for their critical applications?

2. Translate the following into Chinese

carbon-containing gas	electrical	critical application
comprehensive fashion	thermal conductivity	lattice vibration
adjacent aromatic sheet	high-performance carbon fiber	electrical resistance

(1) Pitch-based fibers satisfy the needs of niche markets, and show promise of reducing prices to make mass markets possible.

(2) Projections of the data obtained to ambient temperatures indicate that creep deformations will be infinitesimally small.

(3) This structure appears very rarely in carbon fibers, especially in PAN-based fibers, even though they are conventionally called graphite fibers.

(4) They are the least expensive pitch-based fiber, and are useful in enhancing modulus or conductivity in many applications.

(5) In a comparison of materials for tension members of tension leg platforms for deep sea oil production, carbon fiber strand survived 2000000 stress cycles between 296 and 861 MPa.

3. Translate the following into English

沥青基纤维	芳环结构片层	杨氏模量
烃类气体	应力循环	
大规模生产	蠕变	

(1) 加热后，中间相沥青纤维的模量大幅提升。

(2) 高性能纤维有芳环结构片层构成，这些芳环结构片层彼此随机排列，这种结构称为涡流层状结构。

在线习题

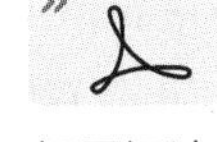
拓展阅读

Unit 2 Glass fibers

Glass fibers are the most common of all reinforcing fibers for polymeric matrix composites (PMC). The principal advantages of glass fibers are low cost, high tensile strength, high chemical resistance, and excellent insulating properties. The disadvantages are relatively low tensile modulus and high density among the commercial fibers, sensitivity to abrasion during handling (which frequently decreases its tensile strength), relatively low fatigue resistance, and high hardness (which causes excessive wear on molding dies and cutting tools).

The two types of glass fibers commonly used in the fiber-reinforced plastics (FRP) industry are E-glass fiber and S-glass fiber. Another type, known as C-glass fiber, is used in chemical applications requiring greater corrosion resistance to acids than is provided by E-glass fibers. E-glass fiber has the lowest cost of all commercially available reinforcing fibers, which is the reason for its widespread use in the FRP industry. S-glass fiber, originally developed for aircraft components and missile casings, has the highest tensile strength among all fibers in use. However, the compositional difference and higher manufacturing cost make it more expensive than E-glass.

The chemical compositions of E-glass fibers and S-glass fibers are shown in Table 3.2. As in common soda-lime glass (window and container glasses), the principal ingredient in all glass fibers is silica (SiO_2). Other oxides, such as B_2O_3 and Al_2O_3, are added to modify the network structure of SiO_2 as well as to improve its workability. Unlike soda-lime glass, the Na_2O and K_2O contents in E-glass fibers and S-glass fibers are quite low, which gives them a better corrosion resistance to water as well as higher surface resistivity. The internal structure of glass fibers is a three-dimensional, long network of silicon, oxygen, and other atoms arranged in a random fashion. Thus, glass fibers are amorphous (noncrystalline) and isotropic (equal properties in all directions).

Table 3.2 Typical compositions of glass fibers (in wt.%)

Type	SiO_2	Al_2O_3	CaO	MgO	B_2O_3	Na_2O
E-glass	54.5	14.5	17	4.5	8.5	0.5
S-glass	64	26	—	10	—	—

The manufacturing process for glass fibers is depicted in the flow diagram in Fig. 3.9. Various ingredients in the glass formulation are first dry-mixed and melted in a refractory furnace at about 1370℃. The molten glass is exuded through a number of orifices contained in a platinum bushing and rapidly drawn into filaments of ～10 μm in diameter. A protective coating (size) is then applied on individual filaments before they are gathered together into a strand and wound on a drum. The coating or size is a mixture of lubricants (which prevent abrasion between the filaments), antistatic agents (which reduce static friction between the filaments), and a binder (which packs the filaments together into a strand). It may also contain small percent ages of a coupling agent that promotes adhesion between fibers and the specific polymer matrix for which it is formulated.

The basic commercial form of continuous glass fiber s is a strand, which is a collection of parallel filaments numbering 204 or more. A roving is a group of untwisted parallel strands (also called ends) wound on a cylindrical forming package. Rovings are used in continuous molding operations, such as filament winding and pultrusion. They can also be preimpregnated with a thin layer of polymeric resin matrix to form prepregs. Prepregs are subsequently cut into required dimensions, stacked, and cured into the final shape in batch molding operations, such as compression molding and hand layup molding.

Chopped strands are produced by cutting continuous strands into short lengths. Chopped strands ranging in length from 3.2 to 12.7 mm are used in injection-molding operations. Longer strands, up to 50.8 mm in length, are mixed with a resinous binder and spread in a two-dimensional random fashion to form copped strand mats (CSMs). These mats are used mostly for hand layup moldings and provide nearly equal properties in all directions in the plane of the structure. Milled glass fibers are produced by grinding continuous strands in a hammer mill into lengths ranging from

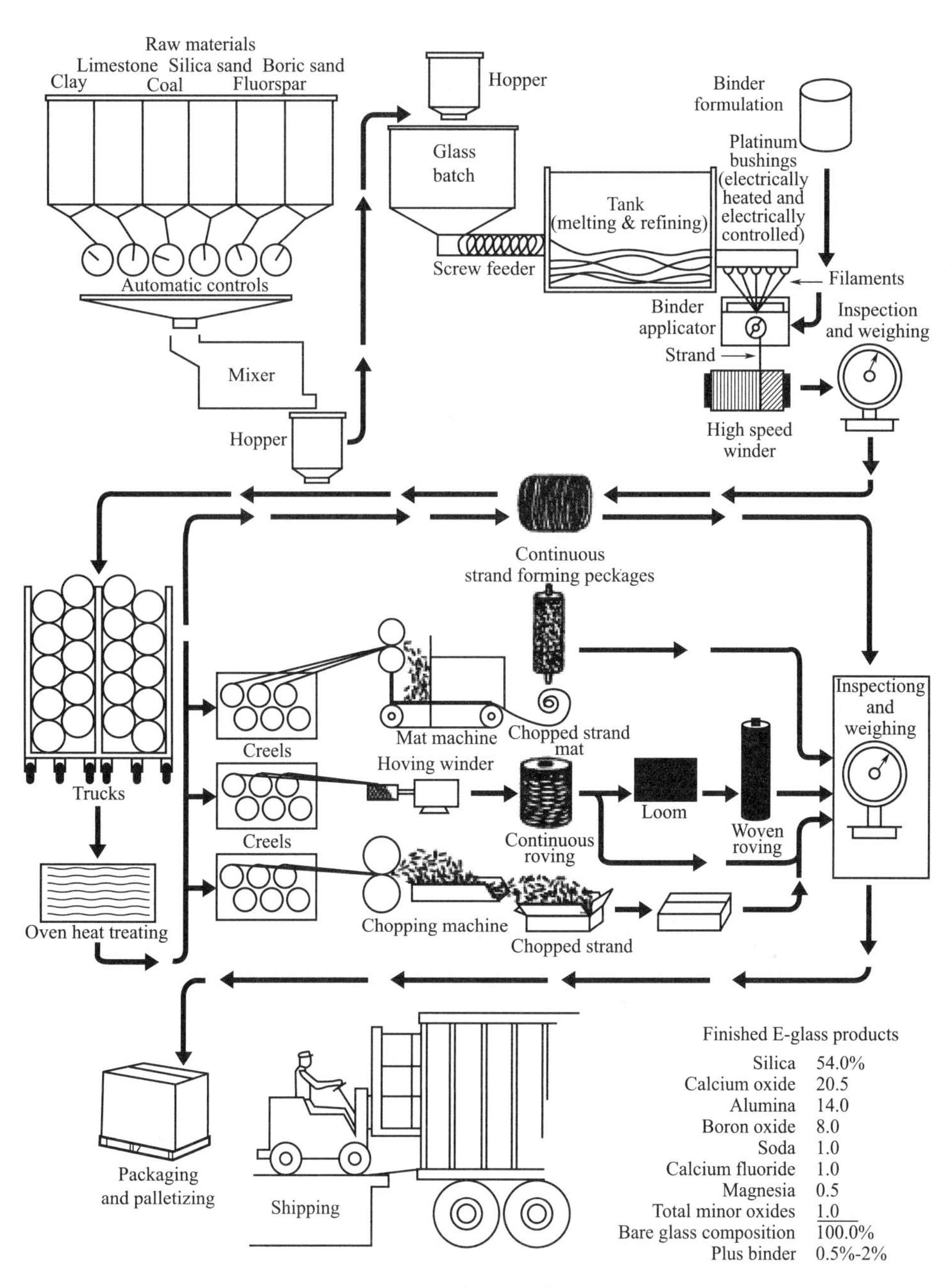

Fig. 3.9 Flow diagram in glass fiber manufacturing

0.79 to 3.2 mm. They are primarily used as a filler in the plastics industry and do not possess any significant reinforcement value. Glass fibers are also available in woven form, such as woven roving or woven cloth. Woven roving is a coarse drapable fabric in which continuous rovings are woven in two mutually perpendicular directions. Woven cloth is weaved using twisted continuous strands, called yarns. Both woven roving and cloth provide bidirectional properties that depend on the style of weaving as well as relative fiber counts in the length (warp) and crosswise (fill) directions. A layer of woven roving is sometimes bonded with a layer of CSM to produce a woven roving mat. All of these forms of glass fibers are suitable for hand layup molding and liquid composite molding.

The average tensile strength of freshly drawn glass fibers may exceed 3.45 GPa. However, surface damage (flaws) produced by abrasion, either by rubbing against each other or by contact with the processing equipment, tends to reduce it to values that are in the range of 1.72-2.07 GPa. Strength degradation is increased as the surface flaws grow under cyclic loads, which is one of the major disadvantages of using glass fibers in fatigue applications. Surface compressive stresses obtained by alkali ion exchange or elimination of surface flaws by chemical etching may reduce the problem; however, commercial glass fibers are not available with any such surface modifications.

The tensile strength of glass fibers is also reduced in the presence of water or under sustained loads (static fatigue). Water bleaches out the alkalis from the surface and deepens the surface flaws already present in fibers. Under sustained loads, the growth of surface flaws is accelerated owing to stress corrosion by atmospheric moisture. As a result, the tensile strength of glass fibers is decreased with increasing time of load duration.

New words and expressions

polymeric *adj*. 聚合的
sensitivity *n*. 敏感
abrasion *n*.（表层）磨损处，磨损
casing *n*. 套，罩
ingredient *n*. 成分，原料
workability *n*. 可使用性，施工性能，可加工性
refractory *n*. 耐火物质
molten *adj*. 熔化的，熔融的
orifices *n*. 孔
platinum *n*. 铂
lubricant *n*. 润滑剂，润滑油
filament *n*. 细丝，丝状物
antistatic *adj*. 抗静电的
static friction 静态摩擦力
parallel *adj*. 平行的
untwist *v*. 解开，未扭曲
preimpregnate *n*. 预浸（渍）
perpendicular *adj*. 垂直的，成直角的
bleach *n*. 漂白剂

Notes

(1) The principal advantages of glass fibers are low cost, high tensile strength, high chemical resistance, and excellent insulating properties.
玻璃纤维的主要优点是成本低、抗拉伸强度高、耐化学性高、绝缘性能优异。

(2) E-glass has the lowest cost of all commercially available reinforcing fibers, which is the reason for its widespread use in the FRP industry.
E 玻璃是所有商用增强纤维中成本最低的，这也是其在纤维增加高分子材料工业中广泛应用的原因。

(3) Other oxides, such as B_2O_3 and Al_2O_3, are added to modify the network structure of SiO_2 as well as to improve its workability.
加入其他氧化物，如 B_2O_3 和 Al_2O_3，以调节 SiO_2 的网络结构并改善其工作性能。

(4) Rovings are used in continuous molding operations, such as filament winding and pultrusion. They can also be preimpregnated with a thin layer of polymeric resin matrix to form prepregs.

粗纤维用于连续成型操作，如长丝缠绕和拉挤。粗纤维也可以预先浸渍一层薄薄的聚合树脂，以形成预浸料。

(5) Strength degradation is increased as the surface damage flaws grow under cyclic loads, which is one of the major disadvantages of using glass fibers in fatigue applications.
在循环载荷作用下，随着表面缺陷的增长，强度退化会增加，这是疲劳应用中使用玻璃纤维的主要缺点之一。

Exercises

1. Questions for discussion

（1）Why glass fibers are the most common of all reinforcing fibers for polymeric matrix composites?

（2）What are the differences between E-glass fiber and S-glass fiber?

（3）Can you talk about the manufacturing process for glass fibers based on the Fig. 3.9?

（4）How can make prepregs from rovings?

2. Translate the following into Chinese

fatigue resistance	aircraft components	manufacturing cost
network structure	individual filaments	hand layup molding
surface modifications	surface flaws	atmospheric moisture

（1）S-glass fiber, originally developed for aircraft components and missile casings, has the highest tensile strength among all fibers in use.

（2）Unlike soda-lime glass, the Na_2O and K_2O content in E- and S-glass fibers is quite low, which gives them a better corrosion resistance to water as well as higher surface resistivity.

（3）The molten glass is exuded through a number of orifices contained in a platinum bushing and rapidly drawn into filaments of ~10 μm in diameter.

（4）Milled glass fibers are produced by grinding continuous strands in a hammer mill into lengths ranging from 0.79 to 3.2 mm.

（5）Both woven roving and cloth provide bidirectional properties that depend on the style of weaving as well as relative fiber counts in the length (warp) and crosswise (fill) directions.

3. Translate the following into English

纤维增强塑料	导弹外壳	钠钙玻璃
耐火炉	静摩擦	两个相互垂直的方向
持续荷载	应力腐蚀	

（1）玻璃配方中的各种成分首先在约 1370℃的耐火炉中进行干燥、混合和熔化。

（2）商业化的连续玻璃纤维是绳束，它是由 204 或更多的平行长丝的集合。

（3）短切原丝是通过将连续的原丝切割成短的长度来生产的。

（4）在持续载荷作用下，由于大气水分的应力腐蚀，表面缺陷的生长速度加快。

4. Scenario simulation

Given you are an engineer from a glass fiber manufacturing plant and the undergraduate students majoring in materials are coming to your company for internship, please introduce the preparation process of glass fiber to them within no more than 5 minutes.

Unit 3 Incorporation of fibers into matrix

Processes for incorporating fibers into a polymer matrix can be divided into two categories. In one category, fibers and matrix are processed directly into the finished product or structure. Examples of such processes are filament winding and pultrusion. In the second category, fibers are incorporated into the matrix to prepare ready-to-mold sheets that can be stored and later processed to form laminated structures by autoclave molding or compression molding. In this section, we briefly describe the processes used in preparing these ready-to-mold sheets. Knowledge of these processes will be helpful in understanding the performance of various composite laminates. Methods for manufacturing composite structures by filament winding, pultrusion, autoclave molding, compression molding, and others are described in the next section.

Ready-to-mold fiber-reinforced polymer sheets are available in two basic forms, prepregs and sheet-molding compounds.

（1）**Prepregs** These are thin sheets of fibers impregnated with predetermined amounts of uniformly distributed polymer matrix. Fibers may be in the form of continuous rovings, mat, or woven fabric. Epoxy is the primary matrix material in prepreg sheets, although other thermoset and thermoplastic polymers have also been used. The width of prepreg sheets may vary from less than 25 mm to over 457 mm. Sheets wider than 457 mm are called broadgoods. The thickness of a ply cured of prepreg sheets is normally in the range of 0.13-0.25 mm. Resin content in commercially available prepregs is between 30% and 45% by weight.

Unidirectional fiber-reinforced epoxy prepregs are manufactured by pulling a row of uniformly spaced (collimated) fibers through a resin bath containing catalyzed epoxy resin dissolved in an appropriate solvent (Fig. 3.10). The solvent is used to control the viscosity of the liquid resin. Fibers preimpregnated with liquid resin are then passed through a chamber in which heat is applied in a controlled manner to advance the curing reaction to the B-stage. At the end of B-staging, the prepreg sheet is backed up with a release film or waxed paper and wound around a take-up roll. The backup material is separated from the prepreg sheet just before it is placed in the mold to manufacture the composite part. The normal shelf life (storage time before molding) for epoxy prepregs is 6-8 days at 23℃; however, it can be prolonged up to 6 months or more if stored at -18℃.

（2）**Sheet-molding compounds** Sheet-molding compounds (SMC) are thin sheets of fibers precompounded with a thermoset resin and are used primarily in compression molding process. Common thermoset resins for SMC sheets are polyesters and vinyl esters. The longer cure time for epoxies has limited their use in SMC.

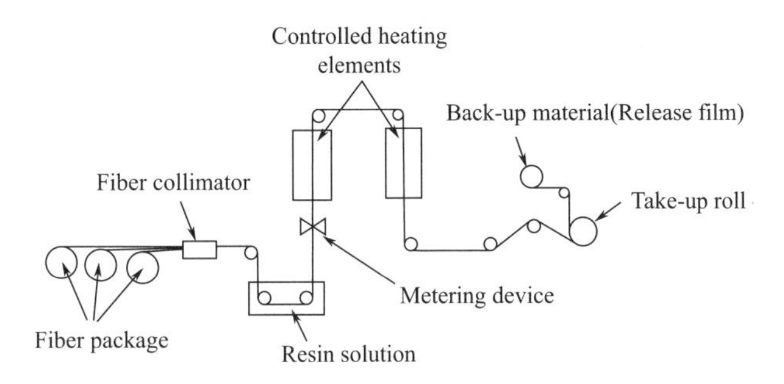

Fig. 3.10 Schematic of prepreg manufacturing

The various types of sheet-molding compounds in current use (Fig. 3.11) are as follows:

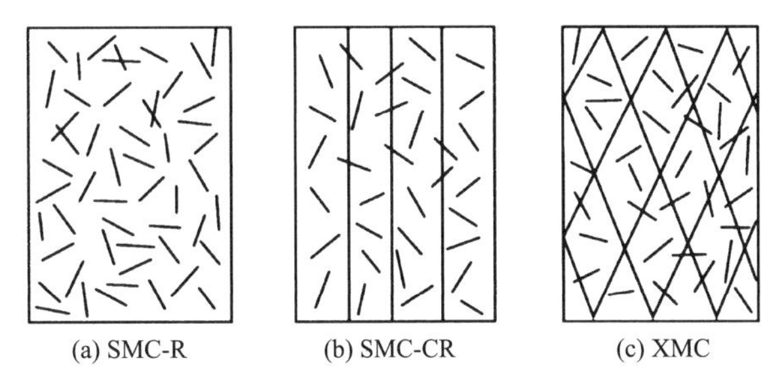

Fig. 3.11 Various types of sheet-molding compounds (SMC)

① **SMC-R**, containing randomly oriented discontinuous fibers. The nominal fiber content (by weight percent) is usually indicated by two-digit numbers after the letter R. For example, the nominal fiber content in SMC-R30 is 30% by weight.

② **SMC-CR**, containing a layer of unidirectional continuous fibers on top of a layer of randomly oriented discontinuous fibers. The nominal fiber contents are usually indicated by two-digit numbers after the letters C and R. For example, the nominal fiber contents in SMC-C40R30 are 40% by weight of unidirectional continuous fibers and 30% by weight of randomly oriented discontinuous fibers.

③ **XMC** (trademark of PPG Industries), containing continuous fibers arranged in an X pattern, where the angle between the interlaced fibers is between 5° and 7°. Additionally, it may also contain randomly oriented discontinuous fibers interspersed with the continuous fibers.

A typical formulation for sheet-molding compound SMC-R30 is presented in Table 3.3. In this formulation, the unsaturated polyester and styrene are polymerized together to form the polyester matrix. The role of the low shrink additive, which is a thermo plastic polymer powder, is to reduce the polymerization shrinkage. The function of the catalyst (also called the initiator) is to initiate the polymerization reaction, but only at an elevated temperature. The function of the inhibitor is to prevent premature curing (gelation) of the resin that may start by the action of the catalyst while the ingredients are blended together. The mold release agent acts as an internal lubricant, and helps in releasing the molded part from the die. Fillers assist in reducing shrinkage of the molded part, promote better fiber distribution during molding, and reduce the overall cost of the compound. Typical filler-resin weight ratios are 1.5:1 for SMC-R30, 0.5:1 for SMC-R50, and nearly 0:1 for

SMC-R65. The thickener is an important component in a SMC formulation since it increases the viscosity of the compound without permanently curing the resin and thereby makes it easier to handle a SMC sheet before molding. However, the thickening reaction should be sufficiently slow to allow proper wet-out and impregnation of fibers with the resin. At the end of the thickening reaction the compound becomes dry, nontacky, and easy to cut and shape. With the application of heat in the mold, the thickening reaction is reversed and the resin paste becomes sufficiently liquid-like to flow in the mold. Common thickeners used in SMC formulations are oxides and hydroxides of magnesium and calcium, such as MgO, $Mg(OH)_2$, CaO, and $Ca(OH)_2$. Another method of thickening is known as the interpenetrating thickening process (ITP), in which a proprietary polyurethane rubber is used to form a temporary three-dimensional network structure with the polyester or vinylester resin.

Table 3.3 Typical formulation of SMC-R30

Material	Weight/%	
Resin paste		
Unsaturated polyester	10.50	
Low shrink additive	3.45	
Styrene monomer	13.40	
Filler ($CaCO_3$)	40.70	70%
Thickener (MgO)	0.70	
Catalyst (TBPB)	0.25	
Mold release agent (zinc stearate)	1.00	
Inhibitor (benzoquinone)	<0.005 g	
Glass fiber (25.4 mm, chopped)		30%
Total		100%

SMC-R and SMC-CR sheets are manufactured on a sheet-molding compound machine (Fig. 3.12). The resin paste is prepared by mechanically blending the various components listed in Table 3.3. It is placed on two moving polyethylene carrier films behind the metering blades. The thickness of the resin paste on each carrier film is determined by the vertical adjustment of the metering blades. Continuous rovings are fed into the chopper arbor, which is commonly set to provide 25.4 mm long discontinuous fibers. Chopped fibers are deposited randomly on the bottom resin paste. For SMC-CR sheets, parallel lines of continuous strand rovings are fed on top of the chopped fiber layer. After covering the fibers with the top resin paste, the carrier films are pulled through a number of compact ion rolls to form a sheet that is then wound around a take-up roll. The wetting of fibers with the resin paste takes place at the compaction stage.

XMC sheets are manufactured by the filament winding process in which continuous strand rovings are pulled through a tank of resin paste and wound under tension around a large rotating cylindrical drum. Chopped fibers, usually 25.4 mm long, are deposited on the continuous fiber layer during the time of winding. After the desired thickness is obtained, the built-up material is cut by a knife along a longitudinal slit on the drum to form the XMC sheet.

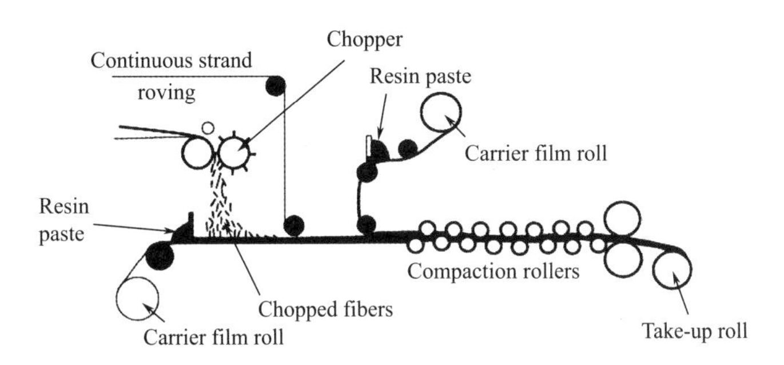

Fig. 3.12 Schematic of a sheet molding compounding operation

At the end of manufacturing, SMC sheets are allowed to "mature" (thicken or increase in viscosity) at about 30℃ for 1-7 days. The matured sheet can be either compression molded immediately or stored at −18℃ for future use.

New words and expressions

incorporate *v.* 将……包括在内，包含
category *n.*（人或事物的）类别
impregnate *v.* 使充满，使遍布，浸渍
predetermined *adj.* 预先确定的，预先决定的
distributed *adj.* 分布的，分散的
woven fabric 纺织品
broadgoods *n.* 宽幅
collimate *v.* 照（对，瞄）准，使成直线
appropriate solvent 合适的溶剂
chamber *n.*（作特定用途的）房间，室
waxed paper *n.*（包装食品或烹饪用的）蜡纸
shelf life *n.* 保质期
precompounded *adj.* 预制的
initiator *n.* 引发剂
inhibitor *n.* 抑制剂
thickener *n.* 增稠剂
premature *adj.* 未成熟的，过早的
gelation *n.* 凝胶化（作用），凝胶的形成（过程）
nontacky *adj.* 非黏性的，不粘的
blade *n.* 刀片；（机器上旋转的）叶片
compact *adj.* 小型的，紧凑的
drum *n.*（尤指机器上的）鼓轮，滚筒

Notes

（1）Fibers are incorporated into the matrix to prepare ready-to-mold sheets that can be stored and later processed to form laminated structures by autoclave molding or compression molding.
将纤维加入基体中，以制备可随时成型的片材，这些片材可被储存起来，随后通过高压釜成型或压缩成型加工成叠层结构。

（2）Unidirectional fiber-reinforced epoxy prepregs are manufactured by pulling a row of uniformly spaced (collimated) fibers through a resin bath containing catalyzed epoxy resin dissolved in an appropriate solvent.
单向纤维增强环氧树脂预浸料是通过将一排均匀间隔（准直）的纤维拉过含有溶解在适当溶剂中的催化环氧树脂的树脂槽而制成的。

（3）The normal shelf life (storage time before molding) for epoxy prepregs is 6-8 days at 23℃; however, it can be prolonged up to 6 months or more if stored at −18 ℃.
环氧预浸料在 23℃下的正常保质期（成型前的储存时间）为 6-8 天；但是，如果在-18℃下储存，则可延长至 6 个月或更长。

（4）Epoxy is the primary matrix material in prepreg sheets, although other thermoset and thermoplastic polymers have also been used.
环氧树脂是预浸料片材的主要基体材料，但也使用了其他热固性和热塑性聚合物。

（5）The nominal fiber content (by weight percent) is usually indicated by two-digit numbers after the letter R. For example, the nominal fiber content in SMC-R30 is 30% by weight.
通常用字母 R 后的两个数字表示标称纤维含量（按重量百分比）。例如，SMC-R30 中的标称纤维含量为 30%（按重量百分比）。

（6）XMC sheets are manufactured by the filament winding process in which continuous strand rovings are pulled through a tank of resin paste and wound under tension around a large rotating cylindrical drum.
XMC 片材是由长丝缠绕工艺生产的，在这种工艺中，连续的无捻粗纱被拉过树脂浆料，然后在一个大的旋转圆柱形滚筒周围张力缠绕。

Exercises

1. Questions for discussion

（1）What is ready-to-mold sheets?
（2）What is the content of resin in commercially available prepregs ?
（3）What kinds of resins are used in sheet-modling compounds?
（4）Why the thickener is an important component in a SMC formulation?

2. Translate the following into Chinese

filament winding
thermoset and thermoplastic polymers
unidirectional continuous fibers
vinylester resin
uniformly distributed polymer matrix
randomly oriented discontinuous fibers
interpenetrating thickening process
chopper arbor

（1）Sheet-molding compounds (SMC) are thin sheets of fibers precompounded with a thermoset resin and are used primarily in compression molding process.
（2）The function of the inhibitor is to prevent premature curing (gelation) of the resin that may start by the action of the catalyst while the ingredients are blended together.
（3）Fillers assist in reducing shrinkage of the molded part, promote better fiber distribution during molding, and reduce the overall cost of the compound.
（4）The thickener is an important component in a SMC formulation since it increases the viscosity of the compound without permanently curing the resin and thereby makes it easier to handle a SMC sheet before molding.

3. Translate the following into English

叠层结构　高压釜成型　压缩成型
不饱和聚酯　苯乙烯　聚合收缩
聚乙烯载体

（1）随时成型的纤维增强聚合物片材有两种基本形式，预浸料和片材成型化合物。
（2）此外，它还可以包含随机定向的不连续纤维，这些纤维散布在连续纤维中。

（3）低收缩添加剂是一种热塑性聚合物粉末，其作用是减少聚合收缩。

（4）各载膜上的树脂浆料厚度由计量叶片的垂直调整决定。

Chapter 4 Interfaces

Unit 1 Types of bonding at the interface

We can define an interface between a reinforcement and a matrix as the bounding surface between the two across which a discontinuity in some parameter occurs. The discontinuity across the interface may be sharp or gradual. Mathematically, interface is a bidimensional region. In practice, we have an interfacial region with a finite thickness. In any event, an interface is the region through which material parameters, such as concentration of an element, crystal structure, atomic registry, elastic modulus, density, coefficient of thermal expansion, etc., change from one side to another. Clearly, a given interface may involve one or more of these items.

The behavior of composite materials is a result of the combined behavior of the following three entities: fiber or the reinforcing element, matrix, fiber/matrix interface.

The reason the interface in a composite is of great importance is that the internal surface area occupied by the interface is quite extensive. It can easily go as high as 3000 cm^2/cm^3 in a composite containing a reasonable fiber volume fraction. We can demonstrate this very easily for a cylindrical fiber in a matrix. The fiber surface area is essentially the same as the interfacial area. Ignoring the fiber ends, one can write the surface-to-volume ratio (S/V) of the fiber as

$$S/V = 2\pi rl/\pi r^2 l = 2/r \quad (4.1)$$

Where, r and l are the fiber radius and length of the fiber, respectively. Thus, the surface area of a fiber of the interfacial area per unit volume increases as r decreases. Clearly, it is important that the fibers should not be weakened by flaws because of an adverse interfacial reaction. Also, the applied load should be effectively transferred from the matrix to the fibers via the interface. Thus, it becomes extremely important to understand the nature of the interface region of any given composite system under a given set of conditions. Specifically, in the case of a fiber reinforced composite material, the interface or more precisely the interfacial zone, consists of near-surface layers of fiber and matrix and any layer(s) of material existing between these surfaces. Wettability of the fiber by the matrix and the type of bonding between the two components constitute the primary considerations. Additionally, one should determine the characteristics of the interface and how they are affected by temperature, diffusion, residual stresses, and so on.

(1) Wettability Various mechanisms can assist or impede adhesion (Baier et al., 1968). A key concept in this regard is that of wettabillity. Wettability tells us about the ability of a liquid to spread on a solid surface. We can measure the wettability of a given solid by a liquid by considering the equilibrium of forces in a system consisting of a drop of liquid resting on a place solid surface in the appropriate atmosphere. Fig. 4.1 shows the situation schematically. The liquid drop will spread and wet the surface completely only if this results in a net reduction of the system free energy. Note that a portion of the solid/vapor interface is substituted by the solid/liquid interface. The contact angle, θ, of a liquid on the solid surface fiber is a convenient and important parameter to characterize wettability. Commonly, the contact angle is measured by putting a sessile drop of the liquid on the flat surface of a solid substrate. The contact angle is obtained from the tangents along three interfaces: solid/liquid, liquid/vapor, and solid/vapor. The contact angle, θ, can be measured directly by a goniometer or calculated by using simple trigonometric relationships involving drop dimensions. In theory, one can using the following expression, called Young's equation:

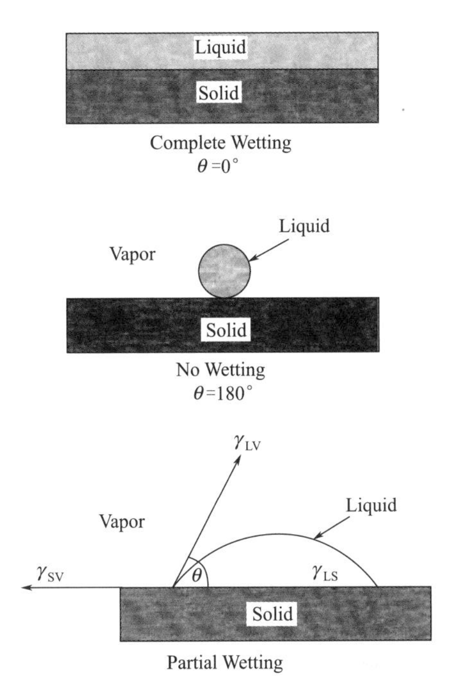

Fig. 4.1 Three difference conditions of wetting: complete wetting, no wetting, and partial wetting. The terms γ_{SV}, γ_{LS} and γ_{LV}, denote the surface energies of solid/vapour, liquid/solid, and liquid/vapor interfaces, respectively

$$\gamma_{SV} = \gamma_{SL} + \gamma_{LV} \cos\theta \tag{4.2}$$

Where, γ is the specific surface energy, and the subscript SV, LS, and LV represent solid/vapor, liquid/solid, and liquid/vapor interfaces, respectively. If this process of substitution of the solid/vapor interface involves an increase in the free energy of the system, then complete spontaneous wetting will not result. Under such conditions, the liquid will spread until a balance of forces acting on the surface is attained; that is, we shall have partial wetting. A small θ implies good wetting. The extreme cases being $\theta = 0°$, corresponding to prefect wetting, and $\theta = 180°$, corresponding to no wetting. In practice, it is rarely possible to obtain a unique equilibrium value of θ. Also, there exists a range of contact angles between the maximum or advancing angle, θ_a, and the minimum or receding angle, θ_r. This phenomenon, called the contact angle hysteresis, is generally observed in polymeric system. Among the sources of this hysteresis are: chemical attack, dissolution, inhomogeneity of chemical composition of solid surface, surface roughness, and local adsorption.

It is important to realize that wettability and bonding are not synonymous system terms. Wettability describes the extent of intimate contact between a liquid and a solid; it does not necessarily mean a strong bond at the interface. One can have excellent wettability and a weak van der Waals-type low-energy bond. A low contact angle, meaning good wettability, is a necessary but not sufficient condition for strong bonding. Consider again a liquid droplet lying on a solid surface. In such a case, Young's equation, Equation 4.2, is commonly used to express the equilibrium among surface tensions in the horizontal direction. What is normally neglected in such an analysis is that there is also a vertical force $\gamma_{LV}\sin\theta$, which must be balanced by a stress in the solid acting perpendicular to the interface. This was first pointed out by Bikerman and Zisman in their discussion of the proof of Young's equation by Johnson (1959). The effect of internal stress in the solid for this configuration was discussed by Cahn et al. (1964, 1979). In general, Young's equation has been applied to void formation in solids without regard to the precise state of internal stress. Fine et al. (1993) analyzed the conditions for occurrence of these internal stresses and their effect on determining work of adhesion in particle reinforced composites.

Wettabilty is very important in PMCs because in the PMC manufacturing the liquid matrix must penetrate and wet fiber tows. Among polymeric resins that are commonly used as matrix materials, thermoset resins have a viscosity in the 1-10 Pa • s range. The melt viscosities of thermoplastics are two to three orders of magnitude higher than those of thermosets and they show, comparatively, poorer fiber wetting characteristics and poorer composites. Although the contact angle is a measure of wettability, the reader should realize that its magnitude will depend on the following important variables: time and temperature of contact; interfacial reactions; stoichiometry, surface roughness and geometry; heat of formation; and electronic configuration.

(2) Types of bonding at the interface It is important to be able to control the degree of bonding between the matrix and the reinforcement. To do so, it is necessary to understand all the different possible bonding types, one or more of which may be acting at any given instant. We can conveniently classify the important types of interfacial bonding as follows: Mechanical bonding, Physical bonding, Chemical bonding, Dissolution bonding, Reaction bonding.

① **Mechanical Bonding** Simple mechanical keying or interlocking effects between two surfaces can lead to a considerable degree of bonding. Any contraction of the matrix onto a central fiber would result in a gripping of the fiber by the matrix. Imagine, for example, a situation in which the matrix in a composite radially shrinks more than the fiber on cooling from a high temperature. This would lead to a gripping of the fiber by the matrix even in the absence of any chemical bonding (Fig. 4.2). The matrix penetrating the crevices on the fiber surface, by liquid or viscous flow or high-temperature diffusion, can also lead to some mechanical bonding. In Fig. 4.2, we have a radial gripping stress, σ_r. This is related to the interfacial shear stress, τ_i, as

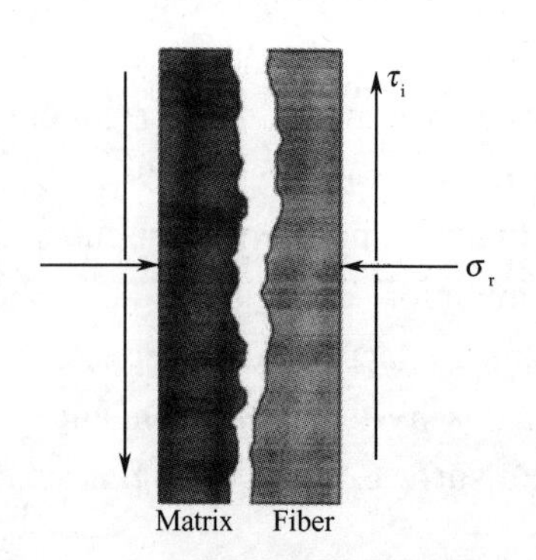

Fig. 4.2 Mechanical gripping due to radial shrinkage of a matrix in a composite more than the fiber on cooling from a high temperature

$$\tau_i = \mu\sigma_r \tag{4.3}$$

Where, μ is the coefficient of friction, generally

between 0.1 and 0.6.

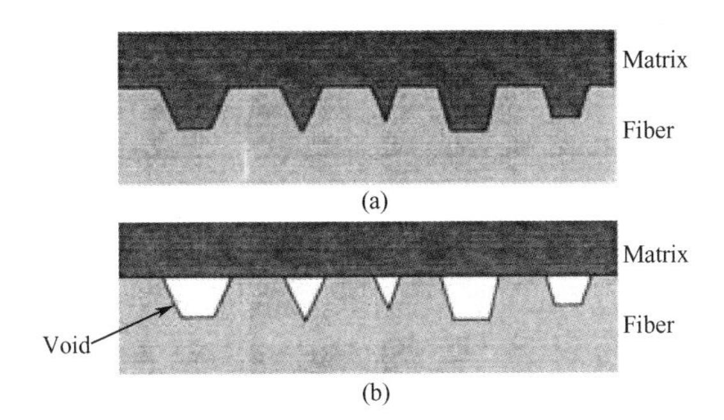

Fig. 4.3 (a) Good mechanical bond (b) Lack of wettability can make a liquid polymer or metal unable to penetrate the asperities on the fiber surface, leading to interfacial voids

In general, mechanical bonding is a low-energy bond vis a vis a chemical bond, i.e., the strength of a mechanical bond is lower than that of a chemical bond. There has been some work (Vennett et al., 1970; Schoene and Scala, 1970) on metallic wires in metal matrices that indicates that in the presence of internal compressive forces, a wetting or metallurgical bond is not quite necessary because the mechanical gripping of the fibers by the matrix is sufficient to cause an effective reinforcement, as indicated by the occurrence of multiple necking in fibers. Hill et al. (1969) confirmed the mechanical bonding effects in tungsten filament/aluminum matrix composites. Chawla and Metzger (1978) studied bonding between an aluminum substrate and anodized alumina (Al_2O_3) films and found that with a rough interface a more efficient load transfer from the aluminum matrix to the alumina occurred. Pure mechanical bonding alone is not enough in most cases. However, mechanical bonding could add, in the presence of reaction bonding, to the overall bonding. Also, mechanical bonding is efficient in load transfer when the applied force is parallel to the interface. In the case of mechanical bonding, the matrix must fill the pores and surface roughness of the reinforcement. Rugosity, or surface roughness, can contribute to bond strength only if the liquid matrix can wet the reinforcement surface. If the matrix, for example, liquid polymer or molten metal, is unable to penetrate the asperities on the fiber surface, then the matrix will solidify and leave interfacial voids, as shown in Fig. 4.3.

We can make some qualitative remarks about general interfacial characteristics that are desirable in different composites. In PMCs and MMCs, one would like to have mechanical bonding in addition to chemical bonding. In CMCs, on the other hand, it would be desirable to have mechanical bonding in lieu of chemical bonding. In any ceramic matrix composite, roughness-induced gripping at the interface is quite important. Specifically, in fiber reinforced ceramic matrix composites, interfacial roughness-induced radial stress will affect the interface debonding, the sliding friction of debonded fibers, and the fiber pullout length.

② **Physical Bonding** Any bonding involving weak, secondary or van der Waals forces, dipolar interactions, and hydrogen bonding can be classified as physical bonding. The bond energy in such physical bonding is approximately 8-16 kJ/mol.

③ **Chemical Bonding** Atomic or molecular transport, by diffusional processes, is involved in chemical bonding. Solid solution and compound formation may occur at the interface, resulting in a reinforcement/matrix interfacial reaction zone with a certain thickness. This encompasses all types of covalent, ionic, and metallic bonding. Chemical bonding involves primary forces and the bond energy in the range of approximately 40-400 kJ/mol.

There are two main types chemical bonding.

a. Dissolution bonding: In this case, interaction between components occurs at an electronic scale. Because these interactions are of rather short range, it is important that the components come into intimate contact on an atomic scale. This implies that surfaces should be appropriately treated to remove any impurities. Any contamination of fiber surfaces, or entrapped air or gas bubbles at the interface, will hinder the required intimate contact between the components.

b. Reaction bonding: In this case, a transport of molecules, atoms, or ions occurs from one or both of the components to the reaction site, that is, the interface. This atomic transport is controlled by diffusional processes. Such a bonding can exist at a variety of interfaces, e.g., glass/polymer, metal/metal, metal/ceramic, or ceramic/ceramic.

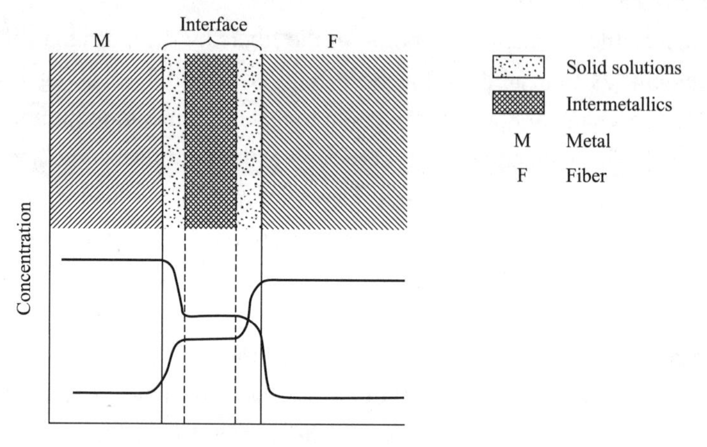

Fig. 4.4 Interface zone in a metal matrix composite showing solid solution and intermetallic compound formation

[*Selected from* Composite materials: Science and Engineering[M]. 2nd ed. Springer, 1998.]

New words and expressions

interface *n.* 界面
bidimensional *adj.* 二维的
elastic modulus 弹性模量
adverse reaction 有害反应，逆反应
interfacial zone 界面区
wettability *n.* 润湿性
residual stress 残余应力
contact angle 接触角
sessile *adj.* 静止的，不动的
goniometer *n.* 角度计，测角器
subscript *n.* 下标
spontaneous *adj.* 自发的
advancing angle 前进角
receding angle 后退角
hysteresis *n.* 滞后
chemical attack 化学腐蚀
inhomogeneity *n.* 非均质性，不均匀性
adsorption *n.* 吸附
roughness *n.*粗糙度
synonymous *adj.* 同义字的，同义的，类义字的
internal stress 内应力
tow *n.* 纤维束，丝束
heat of formation 生成热
stoichiometry *n.* 化学计量学，化学计量关系
interlocking *n.* 咬合
crevice *n.* 裂缝
shear stress 剪切应力
friction *n.* 摩擦
vis a vis 与……相比较，对于，同……相比
alumina *n.* 氧化铝，矾土
anodized *adj.* 受阳极化处理的
necking *n.* 颈缩

metallurgical *adj.* 冶金的，冶金学的，冶金术的
rugosity *n.* 有皱纹，多皱纹性质
asperity *n.* 不平度
in lieu of 代替
void *n.* 空隙
debonding *n.* 脱胶，脱黏
pullout *n.* 拉脱
dipolar *adj.* 偶极的
impurity *n.* 杂质
contamination *n.* 污染

Notes

(1) Thus, the surface area of a fiber of the interfacial area per unit volume increases as r decreases. Clearly, it is important that the fibers should not be weakened by flaws because of an adverse interfacial reaction.
这样，随着 r 的减小，每单位体积的界面面积中的纤维表面积增加。显然，纤维不能因有害的界面反应所产生的瑕疵而削弱，这是非常重要的。

(2) The liquid drop will spread and wet the surface completely only if this results in a net reduction of the system free energy.
只有在可以导致系统的自由能净减少的情况下，液滴将发生铺展和表面的完全润湿。

(3) The extreme cases being $\theta = 0°$, corresponding to prefect wetting, and $\theta = 180°$, corresponding to no wetting.
（两种）极端的情况，当 $\theta = 0°$，对应的是完全润湿的情况，当 $\theta = 180°$ 时，对应（完全）不润湿的情况。

(4) What is normally neglected in such an analysis is that there is also a vertical force $\gamma_{LV} \sin\theta$, which must be balanced by a stress in the solid acting perpendicular to the interface.
在这样的分析中，垂直方向的力 $\gamma_{LV} \sin\theta$，也就是必须与垂直作用在固体界面处的力相平衡的那个力，通常是忽略的。

(5) Although the contact angle is a measure of wettability, the reader should realize that its magnitude will depend on the following important variables: time and temperature of contact; interfacial reactions; stoichiometry, surface roughness and geometry; heat of formation; and electronic configuration.
虽然，接触角是对润湿性的一种量度，但是读者应该认识到其大小取决于以下重要变量：接触时间与温度、界面作用、化学计量关系、表面粗糙度和几何形状、生成热以及电子构型。

Exercises

1. Question for discussion

(1) Why do we say that interface in a composite is of great importance?
(2) Wettability and bonding are not synonymous system terms. Why?
(3) Give the basic types of interfacial bonding.
(4) What is the mechanical bonding? Please give some examples.
(5) What is the chemical bonding? Please give some examples.

2. Translate the following into Chinese

contact angle hysteresis　contact angle　work of adhesion
specific surface area　melt viscosity　solid solution

(1) We can define an interface between a reinforcement and a matrix as the bounding surface between the two across which a discontinuity in some parameter occurs.
(2) An interface is the region through which material parameters, such as concentration of an element, crystal structure, atomic registry, elastic modulus, density, coefficient of thermal expansion, etc., change from one side to another.
(3) In the case of a fiber reinforced composite material, the interface or more precisely the interfacial zone, consists of near-surface layers of fiber and matrix and any layer(s) of material existing between these surfaces.
(4) We can measure the wettability of a given solid by a liquid by considering the equilibrium of forces in a system consisting of a drop of liquid resting on a place solid surface in the appropriate atmosphere.
(5) If this process of substitution of the solid/vapor interface involves an increase in the free energy of the system, then complete spontaneous wetting will not result.
(6) Under such conditions, the liquid will spread until a balance of forces acting on the surface is attained; that is, we shall have partial wetting.
(7) Any bonding involving weak, secondary or van der Waals forces, dipolar interactions, and hydrogen bonding can be classified as physical bonding.
(8) Any bonding involving weak, secondary or van der Waals forces, dipolar interactions, and hydrogen bonding can be classified as physical bonding.

3. Translate the following into English

表面处理　　后退（接触）角　　比表面能
前进（接触）角　　氢键

(1) 复合材料中的界面之所以重要是因为由界面所占有的内表面积是相当巨大的。
(2) 所谓接触角滞后现象，通常可以在聚合物体系中观察到。
(3) 滞后现象的原因主要有化学腐蚀、溶解、固体表面化学组分的不均一性、表面粗糙度和局部的吸附作用。
(4) 润湿和黏合并不是体系的同义语，认识这一点很重要。
(5) 热塑性聚合物树脂的熔融黏度要比热固性聚合物树脂高 2～3 个数量级，相比较之下，热塑性聚合物树脂对纤维的润湿性能较差，形成复合材料较差。
(6) 与化学键合相比，机械嵌合是一种低能量的黏合，也就是说，机械嵌合的强度低于化学键合的强度。
(7) 当存在内部压缩力时，润湿键合或冶金结合并不完全是必要的，因为基体和纤维的机械键合已经足以带来有效的强化作用，就如纤维的多重颈缩出现一样。

4. Scenario simulation

Suppose that you are an editor of the academic magazine Scientific American, you are required to write an abstract for one of the contributions, “Types of bonding at the interface”. In your writing, you should succinctly introduce the entire contribution and address the purpose of the contribution as well as the importance of the bonding at the interface. You should write at least 150 words but no more than 300 words.

Unit 2 Mechanical characterization of interfaces

1. Interfacial Mechanics

There are two main approaches to the identification of parameters characterizing the "strength" of an interface in a composite. The simpler one is to assume that there is some critical stress level which will cause the interface to debond. This could be a normal stress acting transverse to the interface (crack opening, or mode I, loading) or a shear stress acting parallel to the interface (shear, or mode II, loading). In practice, both types of stress may be acting simultaneously (mixed mode loading). It should be recognized that, whereas a crack in a monolithic material will always tend to follow a mode I path, an interface often represents a plane of weakness along which a crack will propagate even if the stress state at the crack tip is heavily mixed mode or pure shear.

One problem with critical stress level tests is that they tend to be initiation-dominated. The measured stress level is usually that at which a crack is observed to propagate in an unstable manner. Such tests are therefore sensitive to the presence of flaws in the interface, at which local stress concentrations become sufficient to trigger crack growth. This initiation dependence tends to be rather marked for many composites, in which the interfaces are often relatively brittle regions. The flaws at which crack initiation occurs may be statistically representative of the structure of the interface, but they can also be created during specimen preparation. Furthermore, there is a danger with some types of test that local stress concentrations can be generated by small misalignments in the loading arrangements. It is certainly common to observe a wide scatter in measured critical stress levels for interfacial debonding. In general, a more reliable measure of interfacial "strength" can be obtained by adopting a fracture mechanics approach. This is based on examining the energetics of crack propagation. A crack will propagate under steady state conditions when the driving force (expressed as an energy release rate per unit area of crack face) equals the fracture energy of the interface, which is a measure of its toughness (or strength). The initiation problem can be eliminated by ensuring that a pre-crack of suitable size is present before testing starts.

It is often a little complex to apply fracture mechanics-based tests to interfaces in composite at least in conventional fibre composites. This is mainly because the geometry associated with a cylindrical surface means that calculation of the driving force for debonding (difference in stored elastic strain energy before and after crack advance) is not simple. Nevertheless, the disadvantages of critical stress measurements should be recognized and it may be noted that fracture mechanics analyses of several interfacial tests applicable to fibre composites have been developed recently. It should also be appreciated that both types of test are susceptible to effects arising from the presence of residual stresses. If the residual stress state is known, then it should be possible to introduce a correction for its effect on the apparent critical stress level or interfacial fracture energy, but in practice this effect is often overlooked or treated in a simplified manner.

Finally, a distinction should be drawn between initial debonding (crack propagation) and subsequent relative displacement of the crack flanks, i.e. frictional sliding. The latter can be of considerable importance in composites, particularly in terms of the energy absorbed during fibre pull-out, which is usually much greater than that associated with the initial debonding event. Moreover, frictional sliding usually takes place under quasi-steady state conditions, so that measured critical shear stresses for this process are expected to be more reliable than estimated critical stresses for debonding.

2. Test Procedures

Many test methods have been developed to interrogate the mechanical response of interfaces within composite materials. The main methods are listed in Tab. 4.1, together with an indication of some of their characteristics and identification of some references giving detailed information about them. Each test focuses either on the behaviour of single fibres or on the collective response of a group of fibres. In some cases, the test can be carried out on conventional composite material, while in others it is necessary to manufacture a special specimen, commonly one containing a single isolated fibre. The table refers exclusively to fibre composites, in which the interfacial geometry is necessarily cylindrical. As a consequence of this, the interfacial stress state during testing is often predominantly mode Ⅱ (shear parallel to the fibre axis). In some of the tests, tensile stress normal to the interface is also imposed, so that mixed mode loading is generated. (The residual stress state is often such that there is a substantial normal stress across the interface, but, since the thermal expansivities of most matrices are greater than those of most fibres, this is usually compressive and hence must be offset before a mode Ⅰ component to the loading can be generated.) It is thus difficult to generate pure mode Ⅰ loading at a fibre/matrix interface (and, even if this were to be done, reliable detection of debonding events would be problematic in the absence of a shear stress). However, for certain types of composite, such as layered systems or coatings, the interface may be planar. Various tests have been developed for such interfaces, several of which allow mixed mode or pure mode Ⅰ loading.

Tab. 4.1 Classification of the tests developed for measurement of interfacial strength in fibre-reinforced composites

Name of test	Entity Tested	Type of Specimen	Parameter Measured	Loading Mode	References
Fibre Pull-out	Single Fibre	Single Embedded Fibre	Critical Stress or Fracture Energy	Ⅱ	(DiFrancia et al.,1996) (Quek,1998) (Zhang et al.,1999) (Liu and Kagawa, 2000)
Fibre Pull-out	Single Fibre	Composite	Critical Stress or Fracture Energy	Ⅱ	(Tandon and Pagano,1998) (Yue et al.,1998) (Kalton et al.,1998)
Fibre Pull-down	Single Fibre	Composite	Critical Stress or Fracture Energy	Ⅱ	(Kalinka et al., 1997) (Kharrat and Chateauminois, 1997)
Fibre Fragmentation	Single Fibre	Single Embedded Fibre	Critical Fibre Aspect Ratio	Ⅱ	(Tripathi and Jones, 1998) (Shia et al., 2000) (Park et al., 2000)
Protrusion (Slice Compression)	Group of Fibres	Composite	Critical Stress	Ⅱ	(Hsueh et al., 1996) (Kagawa and Hsueh, 1999)
Microbond	Single Fibre	Single Embedded Fibre	Critical Stress	Ⅱ	(Day and Young, 1993) (Schuller et al., 1998) (Kessler et al., 1999)
Fibre Bundle Pull-out	Group of Fibres	Composite	Critical Stress or Fracture Energy	Ⅱ	(Domnanovich et al., 1996) (Sakai et al., 1994)
Tensioned Push-out	Single Fibre	Composite	Critical Stress	Ⅰ/Ⅱ	(Watson and Clyne, 1993) (Kalton et al., 1994)
Cruciform transverse Tension	Single Fibre	Single Embedded Fibres	Critical Stress	Ⅰ/Ⅱ	(Majumdar et al.,1998)

[*Selected from* Clyne T W. Encyclopaedia of Materials: Science and Technology：Composites: MMC, CMC, PMC. New York: Elsevier, 2001.]

New words and expressions

critical stress 临界应力
normal stress 正应力，法向应力
crack opening 裂纹张开
simultaneously *adv.* 同时地
crack *n.* 裂纹
monolithic *adj.* 整体的
crack tip 裂纹端
pure shear 纯剪切
local stress 局部应力
crack growth 裂纹扩展，裂纹增长
specimen *n.* 试样，试件，样品
fracture *n.* 断裂，断裂面
crack face 裂纹面
fracture mechanics 断裂力学
crack advance 裂纹进展，裂纹增长
flank *n.* 侧面, 胁, 侧腹，翼
interrogate *v.* 质问, 审问, 讯问

Notes

(1) It should be recognized that, whereas a crack in a monolithic material will always tend to follow a mode Ⅰ path, an interface often represents a plane of weakness along which a crack will propagate even if the stress state at the crack tip is heavily mixed mode or pure shear.
应该认识到，即使在裂纹端的应力状态是重度的混合模式或纯剪切模式，整块材料上的裂纹总是有遵循模式Ⅰ的趋势，（因为）界面常常代表的是一个薄弱的平面，裂纹沿着这一平面扩展。

(2) A crack will propagate under steady state conditions when the driving force (expressed as an energy release rate per unit area of crack face) equals the fracture energy of the interface, which is a measure of its toughness (or strength).
在稳定的状态下，当驱动力（这个可以用单位裂纹面面积的能量释放速率来表达）等于界面的断裂能量 [也就是用来量度材料韧度（或强度）的那个参数] 时，裂纹将扩展。

(3) It is often a little complex to apply fracture mechanics-based tests to interfaces in composite at least in conventional fibre composites. This is mainly because the geometry associated with a cylindrical surface means that calculation of the driving force for debonding (difference in stored elastic strain energy before and after crack advance) is not simple.
把基于断裂力学原理的检测用于复合材料界面，通常有点复杂，至少对常规的纤维复合材料是这样。这主要是因为与圆柱的表面相关的几何，（这）意味着脱胶驱动力的计算（裂纹增长前后的弹性变形储能的差值）并不简单。

(4) Finally, a distinction should be drawn between initial debonding (crack propagation) and subsequent relative displacement of the crack flanks, i.e frictional sliding. The latter can be of considerable importance in composites, particularly in terms of the energy absorbed during fibre pull-out, which is usually much greater than that associated with the initial debonding event.
最后，应该把初始脱胶与随后的裂纹侧面的相对位移，也就是摩擦滑移，区别开来。后者在复合材料中是相当重要的，特别是从纤维拔脱中吸收能量的角度来看，（因为）拔脱

吸收的能量通常要比初始脱胶吸收的能量要高很多。

(5) In some of the tests, tensile stress normal to the interface is also imposed, so that mixed mode loading is generated. (The residual stress state is often such that there is a substantial normal stress across the interface, but, since the thermal expansivities of most matrices are greater than those of most fibres, this is usually compressive and hence must be offset before a mode I component to the loading can be generated.)
在一些测试中，也施加垂直于界面的拉伸应力，所以会生成混合模式的荷载 [残余应力状态经常是这样的情况：在界面处有一个大的法向应力，但是，由于绝大多数基体的热胀系数比纤维的热胀系数要大，（这样在界面处）会有一个压力（的作用），因此在模式Ⅰ荷载形成前，法向（拉伸）应力会被抵消（一部分）]。注：这里在 hence 之前省略了 normal stress。

Exercises

1. Question for discussion

(1) There are two types of stress that can be used to characterize the strength of an interface in a composite. What are they?
(2) What is frictional sliding? Why is it important in composites?
(3) There are some precautions to be taken in testing the strength of an interface in a composite. Can you give some examples?

2. Translate the following into Chinese

trigger crack growth　　crack tip　　residual stress　　a single isolated fibre

(1) It is certainly common to observe a wide scatter in measured critical stress levels for interfacial debonding.
(2) There are two main approaches to the identification of parameters characterizing the “strength” of an interface in a composite.
(3) It should also be appreciated that both types of test are susceptible to effects arising from the presence of residual stresses.
(4) The flaws at which crack initiation occurs may be statistically representative of the structure of the interface, but they can also be created during specimen preparation.
(5) Each test focuses either on the behaviour of single fibres or on the collective response of a group of fibres.

3. Translate the following into English

临界应力水平　　局部应力集中　　准静态条件
试样制备　　纤维拔脱

(1) 如果残余应力状态已知，那么应该有可能引入一个其对临界应力水平或界面断裂能量影响的修正，但在实践中，这个影响经常被忽略或被简化处理。
(2) 对很多复合材料来说，萌生的依赖性往往非常显著，这些复合材料的界面通常是相对脆弱的区域。

(3) 摩擦滑移通常发生在准静态条件下，因此预计该过程中测量的临界剪切应力比脱粘过程中估计的临界应力更可靠。
(4) 可以发生裂纹萌生的缺陷，从统计学角度上代表了界面的结构，但是，缺陷也可以在试样制备的过程中产生。
(5) 这些测试对界面中瑕疵的存在非常敏感，在有瑕疵存在的条件下，局部应力集中足以激发裂纹增长。

4. Scenario simulation

Suppose that you are an analyst of Analysis and Test Center of Composite Material of Beijing, you have decided to carry out some tests for characterizing the fracture energy of the samples delivered from a composites manufacturing Co. Design an appropriate table listing the method, entity tested, type of specimen and loading mode. You should use the proper terms provide by this text.

在线习题
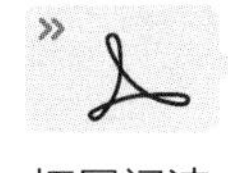
拓展阅读

Chapter 5 Fibre reinforced polymers

扫码听音频

Unit 1 Introduction

In its most basic form a composite material is one which is composed of at least two elements working together to produce material properties that are different to the properties of those elements on their own. In practice, most composites consist of a bulk material (the "matrix"), and a reinforcement of some kind, added primarily to increase the strength and stiffness of the matrix. This reinforcement is usually in fibre form. In the most common man-made composites, Polymer Matrix Composites (PMC's) are the most common and will be discussed here. Also known as FRP - Fibre Reinforced Polymers (or Plastics) – these materials use a polymer-based resin as the matrix, and a variety of fibres such as glass, carbon and aramid as the reinforcement.

Resin systems such as epoxies and polyesters have limited use for the manufacture of structures on their own, since their mechanical properties are not very high when compared to, for example, most metals. However, they have desirable properties, most notably their ability to be easily formed into complex shapes. Materials such as glass, aramid and boron have extremely high tensile and compressive strength but in 'solid form' these properties are not readily apparent. This is due to the fact that when stressed, random surface flaws will cause each material to crack and fail well below its theoretical 'breaking point'. To overcome this problem, the material is produced in fibre form, so that, although the same number of random flaws will occur, they will be restricted to a small number of fibres with the remainder exhibiting the material's theoretical strength. Therefore a bundle of fibres will reflect more accurately the optimum performance of the material. However, fibres alone can only exhibit tensile properties along the fibre's length, in the same way as fibres in a rope.

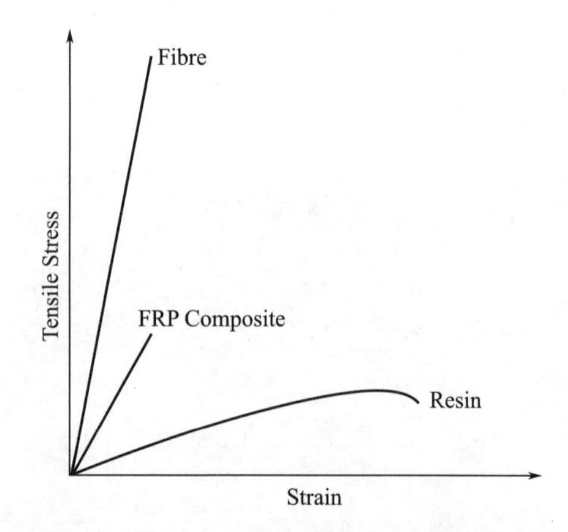

Fig. 5.1 Schematic tensile stress-strain diagram of fibre, FRP composite and resin

It is when the resin systems are combined with reinforcing fibres such as glass, carbon and aramid, that exceptional properties can be obtained. The resin matrix spreads the load applied to the composite between each of the individual fibres and also protects the fibres from damage caused by abrasion and impact. High strengths and stiffnesses, ease of moulding complex shapes, high environmental resistance all coupled with low densities, make the resultant composite superior to metals for many applications.

Since PMC's combine a resin system and reinforcing fibres, the properties of the resulting composite material will combine something of the properties of the resin on its own with that of the fibres on their own (see Fig. 5.1).

Overall, the properties of the composite are determined by:
① The properties of the fibre;
② The properties of the resin;
③ The ratio of fibre to resin in the composite (Fibre Volume Fraction);
④ The geometry and orientation of the fibres in the composite.

The first two will be dealt with in more detail later. The ratio of the fibre to resin derives largely from the manufacturing process used to combine resin with fibre, as will be described in the section on manufacturing processes. However, it is also influenced by the type of resin system used, and the form in which the fibres are incorporated. In general, since the mechanical properties of fibres are much higher than those of resins, the higher the fibre volume fraction the higher will be the mechanical properties of the resultant composite. In practice there are limits to this, since the fibres need to be fully coated in resin to be effective, and there will be an optimum packing of the generally circular cross-section fibres. In addition, the manufacturing process used to combine fibre with resin leads to varying amounts of imperfections and air inclusions.

Typically, with a common hand lay-up process as widely used in the boat-building industry, a limit for FVF is approximately 30%-40%. With the higher quality, more sophisticated and precise processes used in the aerospace industry, FVF's approaching 70% can be successfully obtained.

The geometry of the fibres in a composite is also important since fibres have their highest mechanical properties along their lengths, rather than across their widths. This leads to the highly anisotropic properties of composites, where, unlike metals, the mechanical properties of the composite are likely to be very different when tested in different directions. This means that it is very important when considering the use of composites to understand at the design stage, both the magnitude and the direction of the applied loads. When correctly accounted for, these anisotropic properties can be very advantageous since it is only necessary to put material where loads will be applied, and thus redundant material is avoided.

It is also important to note that with metals the properties of the materials are largely determined by the material supplier, and the person who fabricates the materials into a finished structure can do almost nothing to change those 'in-built' properties. However, a composite material is formed at the same time as the structure is itself being fabricated. This means that the person who is making the structure is creating the properties of the resultant composite material, and so the manufacturing processes they use have an unusually critical part to play in determining the performance of the resultant structure.

There are four main direct loads that any material in a structure has to withstand: tension, compression, shear and flexure.

(1) Tension Fig. 5.2 shows a tensile load applied to a composite. The response of a composite to tensile loads is very dependent on the tensile stiffness and strength properties of the reinforcement fibres, since these are far higher than the resin system on its own.

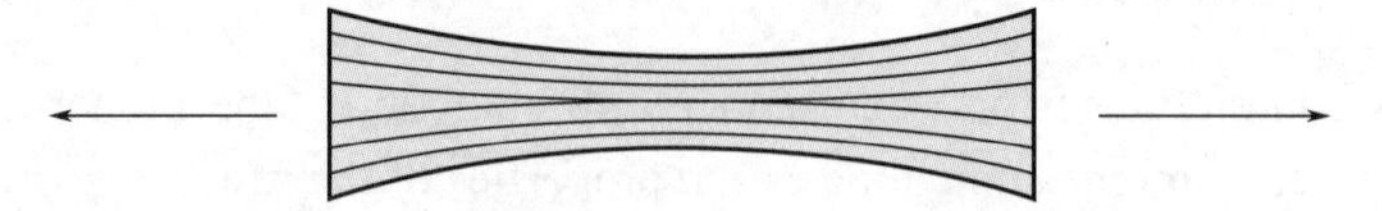

Fig. 5.2 Tensile load applied to a composite

(2) Compression Fig. 5.3 shows a composite under a compressive load. Here, the adhesive and stiffness properties of the resin system are crucial, as it is the role of the resin to maintain the fibres as straight columns and to prevent them from buckling.

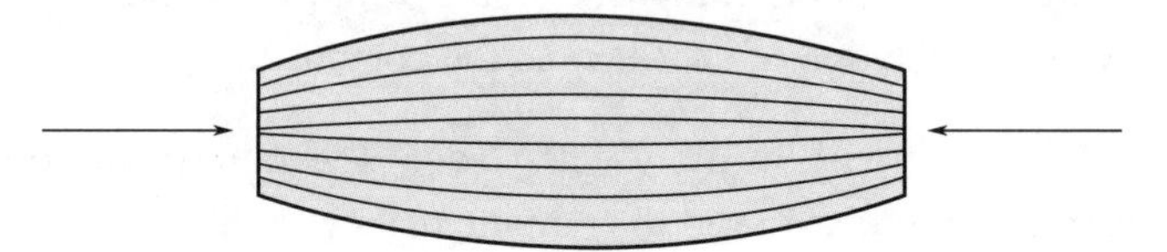

Fig. 5.3 Composite under a compressive load

(3) Shear Fig. 5.4 shows a composite experiencing a shear load. This load is trying to slide adjacent layers of fibres over each other. Under shear loads the resin plays the major role, transferring the stresses across the composite. For the composite to perform well under shear loads the resin element must not only exhibit good mechanical properties but must also have high adhesion to the reinforcement fibre. The interlaminar shear strength of a composite is often used to indicate this property in a multilayer composite ("laminate").

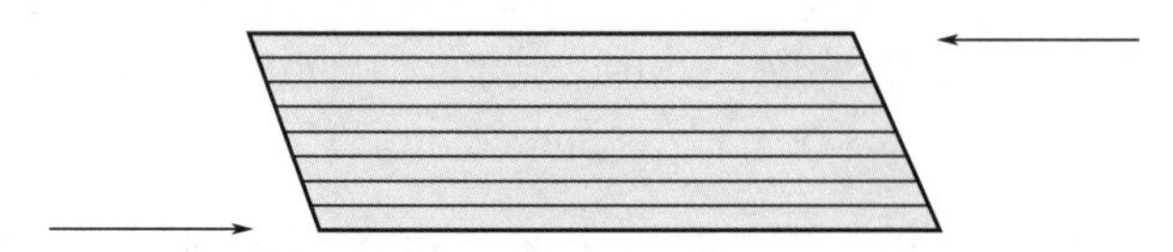

Fig. 5.4 Composite experiencing a shear load

(4) Flexure Flexural loads are really a combination of tensile, compression and shear loads. When loaded as shown, the upper face is put into compression, the lower face into tension and the central portion of the laminate experiences shear (Fig. 5.5).

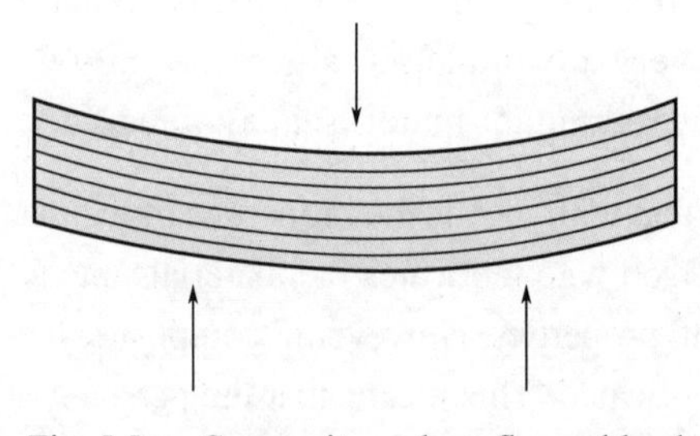

Fig. 5.5 Composite under a flexural load

New words and expressions

bulk *n.* 体积，本体

resin *n.* 树脂

aramid *n.* 聚芳香酰胺类

epoxies *n.* 环氧基

boron *n.* 硼
tensile *adj.* 拉长的，拉力的，拉伸的
compressive *adj.* 压缩的
stress *n.* 应力
optimum *adj.* 最适，最优
abrasion *n.* 磨损，磨光
impact *n.* 冲击
moulding *n.* 模塑
geometry *n.* 几何学
orientation *n.* 定向，方位
incorporated *adj.* 合成一体的
FVF(fiber volume fractures) 纤维体积含量
sophisticated *adj.* 复杂的，尖端的，高级的
aerospace *n.* 航天
anisotropic *adj.* 非等方向的，各向异性的
compression *n.* 压缩，浓缩
slide *v.* 滑动，滑行，滑道，滑梯
interlaminar *adj.* 层间
laminate *v.* 制成薄板，制成箔，形成薄板；*adj.* 由薄片组成的，薄板状的
flexure *n.* 弯曲；拐度，曲率；弯曲部分；单斜挠褶
hand lay-up 手糊工艺

Notes

(1) In practice, most composites consist of a bulk material (the 'matrix'), and a reinforcement of some kind, added primarily to increase the strength and stiffness of the matrix.
在实践中，大多数复合材料包含基体材料和增强材料，增强材料的主要目的是为了增加基体的强度和刚度。

(2) Since PMC's combine a resin system and reinforcing fibres, the properties of the resulting composite material will combine something of the properties of the resin on its own with that of the fibres on their own.
由于 PMC 结合了树脂和增强纤维，所产生的复合材料的性能会结合增强纤维与树脂的某些性能。

(3) The geometry of the fibres in a composite is also important since fibres have their highest mechanical properties along their lengths, rather than across their widths.
复合材料中纤维的几何形状也很重要，因为纤维沿其长度方向有最高的力学性能，而不是在宽度方向。

Exercises

1. Question for discussion

(1) What are Polymer Matrix Composites?
(2) What are the main components of polymer matrix composite?
(3) Give some examples of the reinforcing fibers.
(4) Why are the properties of composite material different to the properties of components?
(5) Why the properties are not readily apparent in solid form?
(6) What is the role of random flaw?
(7) What is the role of resin matrix in composite material?
(8) What is the reason that resin systems have limited use for the manufacture of structure on their own?
(9) What is the reason that fibers will reflect more accurately the optimum performance of the material?
(10) What factors do influence the properties of composite materials?
(11) What is the most notably property of resin?

(12) What is the advantage of FRP?
(13) What is the percentage of fiber in aerospace?

2. Translate the following into Chinese

bulk material	geometry and orientation	compression
polymer Matrix Composites	aerospace industry	shear
fiber Reinforced Polymers	magnitude	flexure
mechanical properties	redundant material	stiffness properties
manufacturing processes	‘in-built’ properties	shear loads
anisotropic properties	tension	interlaminar

(1) Resin systems such as epoxies and polyesters have limited use for the manufacture of structures on their own, since their mechanical properties are not very high when compared to, for example, most metals.
(2) The resin matrix spreads the load applied to the composite between each of the individual fibers and also protects the fibers from damage caused by abrasion and impact.
(3) In general, since the mechanical properties of fibers are much higher than those of resins, the higher the fiber volume fraction the higher will be the mechanical properties of the resultant composite.
(4) These materials use a polymer-based resin as the matrix, and a variety of fibers such as glass, carbon and aramid as the reinforcement.
(5) Materials such as glass, aramid and boron have extremely high tensile and compressive strength but in ‘solid form’ these properties are not readily apparent.
(6) This is due to the fact that when stressed, random surface flaws will cause each material to crack and fail well below its theoretical ‘breaking point’.
(7) The resin matrix spreads the load applied to the composite between each of the individual fibers and also protects the fibers from damage caused by abrasion and impact.
(8) High strengths and stiffnesses, ease of moulding complex shapes, high environmental resistance all coupled with low densities, make the resultant composite superior to metals for many applications.
(9) Since PMC’s combine a resin system and reinforcing fibers, the properties of the resulting composite material will combine something of the properties of the resin on its own with that of the fibers on their own.
(10) The ratio of the fiber to resin derives largely from the manufacturing process used to combine resin with fiber, as will be described in the section on manufacturing processes.
(11) In addition, the manufacturing process used to combine fiber with resin leads to varying amounts of imperfections and air inclusions.
(12) With the higher quality, more sophisticated and precise processes used in the aerospace industry, FVF’s approaching 70% can be successfully obtained.
(13) The geometry of the fibers in a composite is also important since fibers have their highest mechanical properties along their lengths, rather than across their widths.
(14) This means that it is very important when considering the use of composites to understand at the design stage, both the magnitude and the direction of the applied loads.

(15) This means that the person who is making the structure is creating the properties of the resultant composite material, and so the manufacturing processes they use have an unusually critical part to play in determining the performance of the resultant structure.

3. Translate the following into English

力学性质　　　　各向异性性能

层间剪切强度　　　　加工过程

(1) 对剪切载荷下表现良好的复合材料，树脂成分必须不仅展现好的力学性质，而且必须对增强纤维具有高的黏合力。

(2) 复合材料对拉伸载荷的响应强烈依赖于增强纤维的拉伸刚度和拉伸强度。

(3) 玻璃，芳香族聚酰胺和硼等材料具有相当高的拉伸和压缩强度，但是在固体形态，这些性质并不明显。

Unit 2　Fabrication processes

The mixture of reinforcement/resin does not really become a composite material until the last phase of the fabrication, that is, when the matrix is hardened. After this phase, it would be impossible to modify the material, as in the way one would like to modify the structure of a metal alloy using heat treatment, for example. In the case of polymer matrix composites, this has to be polymerized, for example, polyester resin. During the solidification process, it passes from the liquid state to the solid state by copolymerization with a monomer that is mixed with the resin. The phenomenon leads to hardening. This can be done using either a chemical (accelerator) or heat. The following pages will describe the principal processes for the formation of composite parts.

1. Molding processes

The flow chart in Fig. 5.6 shows the steps found in all molding processes. Forming by molding processes varies depending on the nature of the part, the number of parts, and the cost. The mold material can be made of metal, polymer, wood, or plaster.

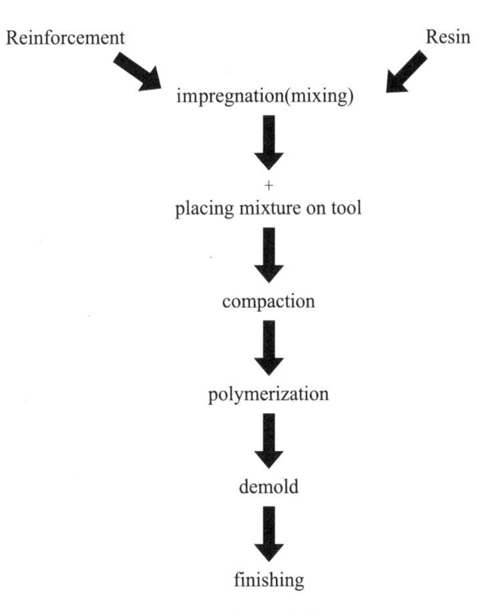

Fig. 5.6　Steps in molding process

(1) Contact molding　Contact molding (see Fig. 5.7) is open molding (there is only one mold, either male or female). The layers of fibers impregnated with resin (and accelerator) are placed on the mold. Compaction is done using a roller to squeeze out the air pockets. The duration for resin setting varies, depending on the amount of accelerator, from a few minutes to a few hours. One can also obtain parts of large dimensions at the rate of about 2 to 4 parts per day per mold.

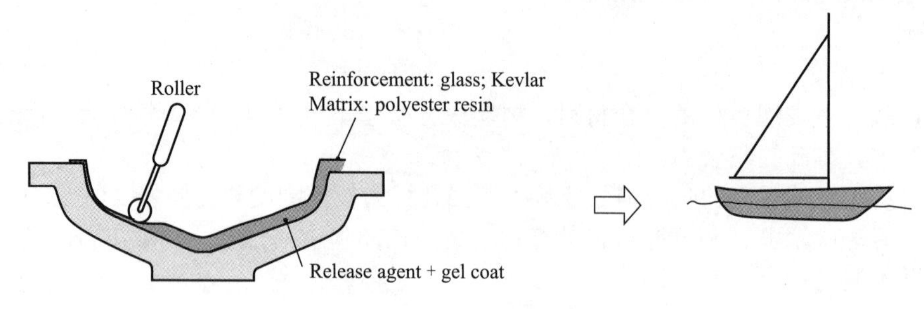

Fig. 5.7 Contact molding

(2) Compression molding With compression molding (see Fig. 5.8), the countermold will close the mold after the impregnated reinforcements have been placed on the mold. The whole assembly is placed in a press that can apply a pressure of 1 to 2 bars. The polymerization takes place either at ambient temperature or higher. The process is good for average volume production: one can obtain several dozen parts a day (up to 200 with heating). This has application for automotive and aerospace parts.

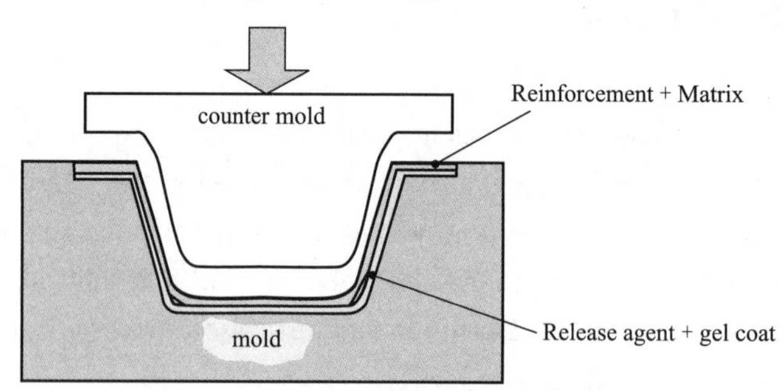

Fig. 5.8 Compression molding

(3) Molding with vacuum This process of molding with vacuum is still called depression molding or bag molding. As in the case of contact molding described previously, one uses an open mold on top of which the impregnated reinforcements are placed. In the case of sandwich materials, the cores are also used. One sheet of soft plastic is used for sealing (this is adhesively bonded to the perimeter of the mold). Vacuum is applied under the piece of plastic (see Fig. 5.9). The piece is then compacted due to the action of atmospheric pressure, and the air bubbles are eliminated. Porous fabrics absorb excess resin. The whole material is polymerized by an oven or by an autoclave under pressure (7 bars in the case of carbon/epoxy to obtain better mechanical properties), or with heat, or with electron beam, or X-rays; see Fig. 5.10). This process has applications for aircraft structures, with the rate of a few parts per day (2 to 4).

(4) Resin injection molding With resin injection molding (see Fig. 5.11), the reinforcements (mats, fabrics) are put in place between the mold and counter mold. The resin (polyester or phenolic) is injected. The mold pressure is low. This process can produce up to 30 pieces per day. The investment is less costly and has application in automobile bodies.

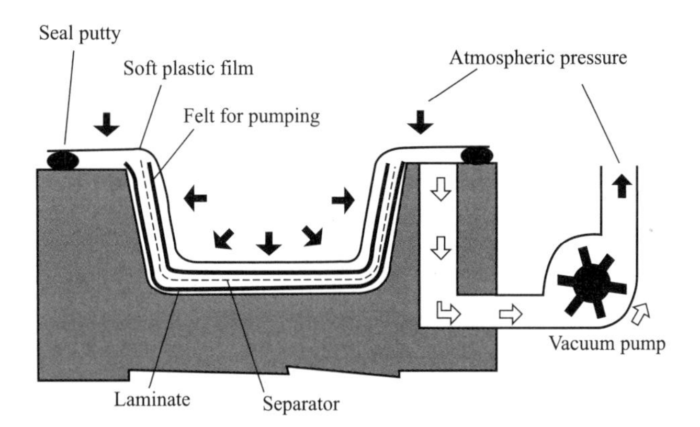

Fig. 5.9 Vaccum molding

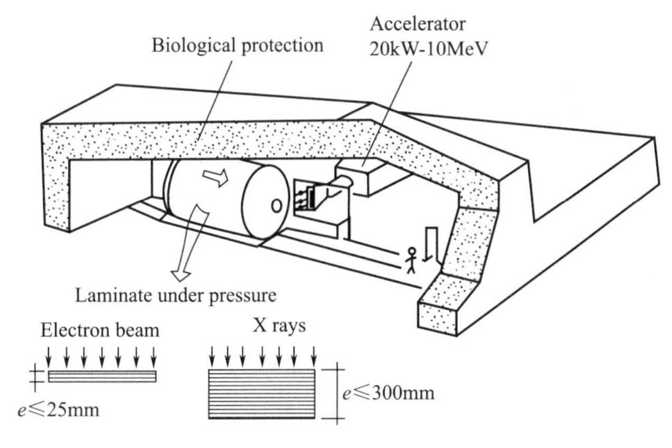

Fig. 5.10 Electron beam of X-ray molding

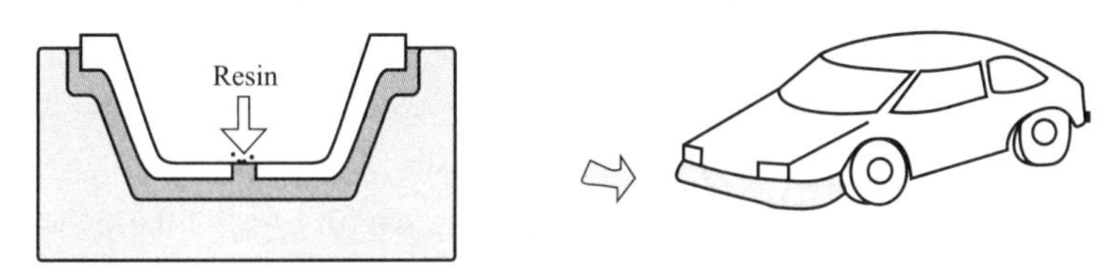

Fig. 5.11 Resin injection molding

(5) Molding by injection of premixed The process of molding by injection of premixed allows automation of the fabrication cycle (rate of production up to 300 pieces per day).

① **Thermoset resins:** Can be used to make components of auto body. The schematic of the process is shown in Fig. 5.12.

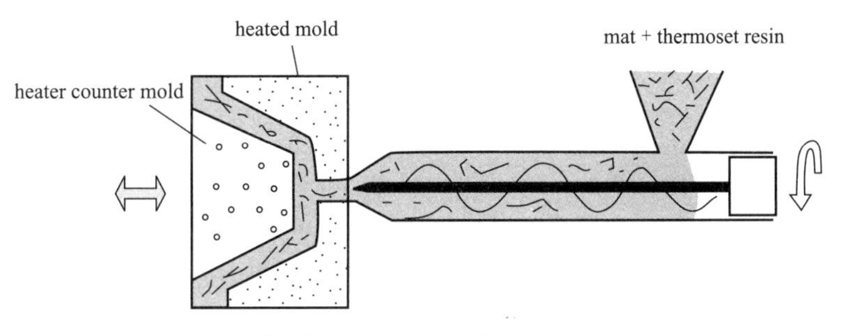

Fig. 5.12 Injection of premixed

② **Thermoplastic resins**: Can be used to make mechanical components with high temperature resistance, as shown in Fig. 5.13.

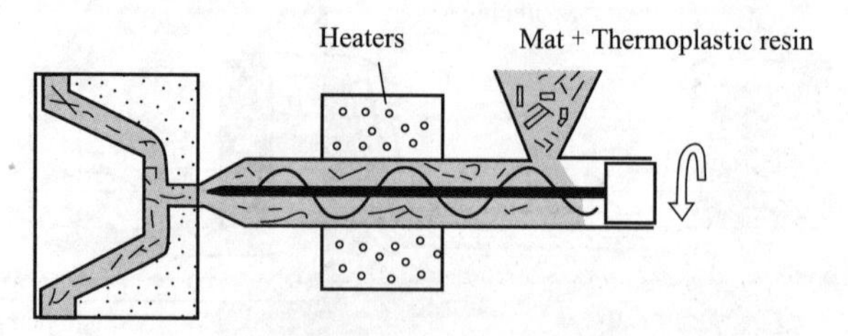

Fig. 5.13 Injection of thermoplastic premixed

(6) Molding by foam injection Molding by foam injection (see Fig. 5.14) allows the processing of pieces of fairly large dimensions made of polyurethane foam reinforced with glass fibers. These pieces remain stable over time, with good surface conditions, and have satisfactory mechanical and thermal properties.

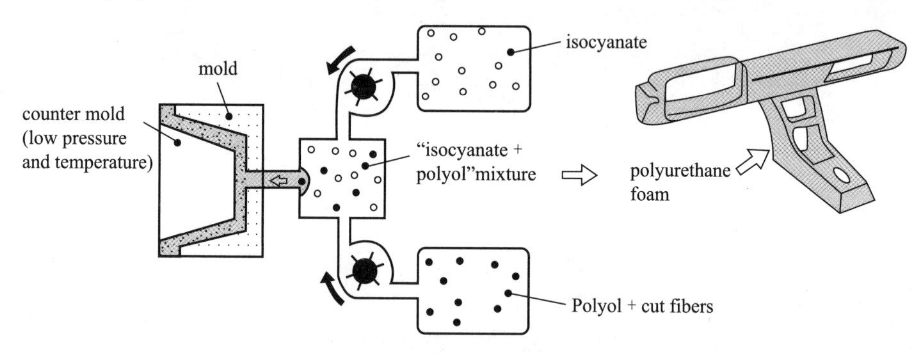

Fig. 5.14 Foam injection

(7) Molding of components of revolution The process of centrifugal molding (see Fig. 5.15) is used for the fabrication of tubes. It allows homogeneous distribution of resin with good surface conditions, including the internal surface of the tube. The length of the tube depends on the length of the mold. Rate of production varies with the diameter and length of the tubes (up to 500 kg of composite per day).

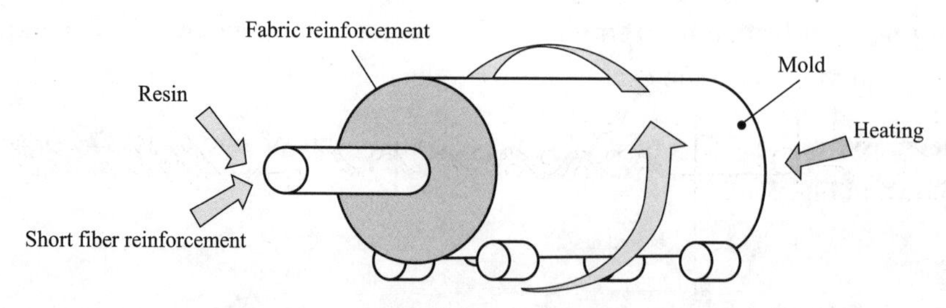

Fig. 5.15 Centrifugal molding

The process of filament winding (see Fig. 5.16) can be integrated into a continuous chain of production and can fabricate tubes of long length. The rate of production can be up to 500 kg of composite per day. These can be used to make missile tubes, torpillas, containers, or tubes for transporting petroleum. For pieces which must revolve around their midpoint, winding can be done on a mandrel. This can then be removed and cured in an autoclave (see Fig. 5.17). The fiber volume fraction is high (up to 85%). This process is used to fabricate components of high internal pressure, such as reservoirs and propulsion nozzles.

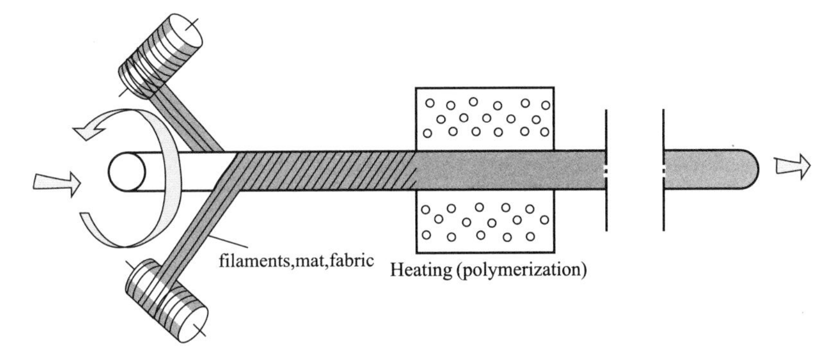

Fig. 5.16 Filament winding

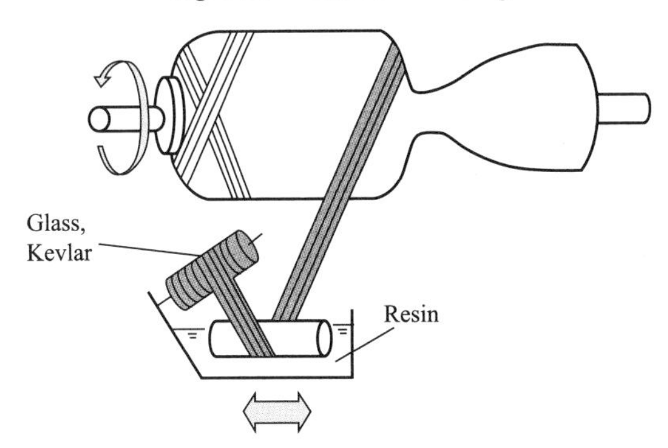

Fig. 5.17 Filament winding on complex mandrel

2. Other forming processes

(1) Sheet forming This procedure of sheet forming (see Fig. 5.18) allows the production of plane or corrugated sheets by corrugation or ribs.

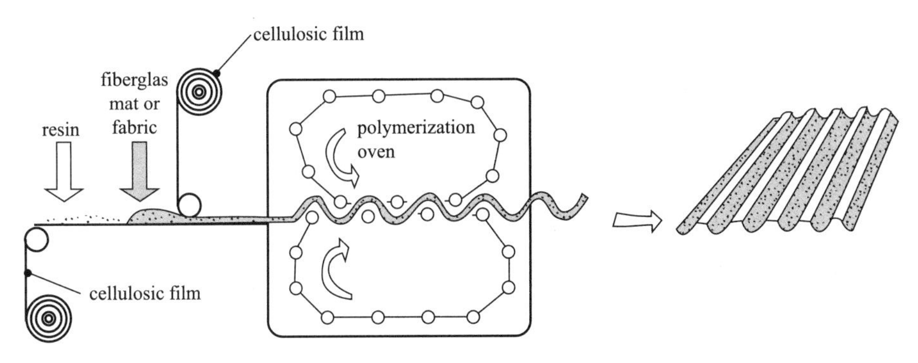

Fig. 5.18 Sheet forming

(2) Profile forming The piece shown in Fig. 5.19 is made by pultrusion. This process makes possible the fabrication of continuous open or closed profiles. The fiber content is important for high mechanical properties. The rate of production varies between 0.5 and 3 m/minute, depending on the nature of the profile.

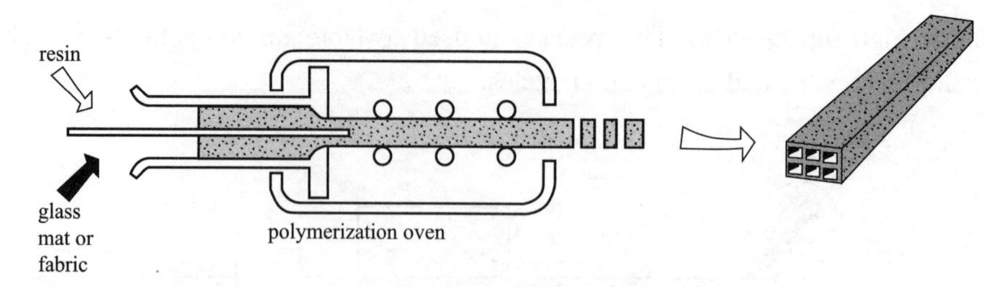

Fig. 5.19 Profile forming

(3) Stamp forming Stamp forming (see Fig. 5.20) is only applicable to thermoplastic composites. One uses preformed plates, which are heated, stamped, and then cooled down.

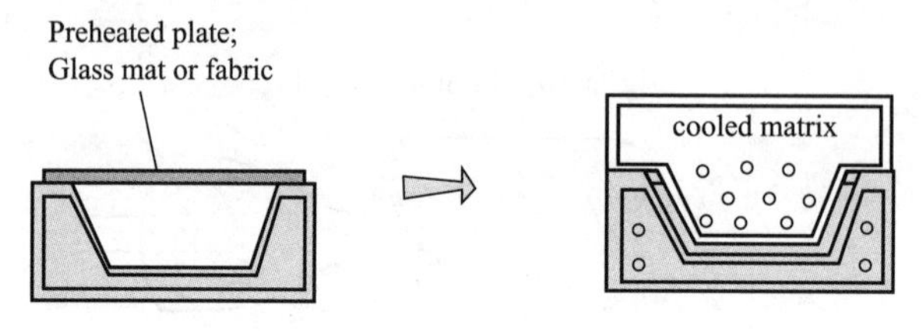

Fig. 5.20 Stamp forming

New words and expressions

polyester resin 聚酯树脂
solidification *n.* 凝固,固化
copolymerization *n.* 共聚合
molding process 成型加工
contact molding 接触模压成型，接触模塑
compression molding 模压成型
impregnated *adj.* 浸渍的
compaction *n.* 真空压实，压实
roller *n.* 滚筒，滚轴
squeeze out 挤出，榨出
depression molding 减压制模
bag molding 袋压成型，袋模成型
porous *adj.* 多孔的
electron beam 电子束，阴极射线
X-ray X射线
phenolic *n.* 酚醛树脂，酚醛塑料
thermoset resin 热固性树脂
thermoplastic resin 热塑性树脂
polyurethane *n.* 聚氨酯（树脂），聚氨基甲酸树脂
mandrel *n.* 芯轴，芯模，芯管，芯棒
propulsion nozzle *n.* 推进喷嘴
corrugated sheet 瓦垅薄波纹板
stamp *v.* 冲压
impregnation *n.* 浸胶
orientable *adj.* 可定向的

Notes

(1) During the solidification process, it passes from the liquid state to the solid state by copolymerization with a monomer that is mixed with the resin.
在固化过程中，通过与混在树脂中单体的共聚合，它从液态变为固态。

(2) Contact molding is open molding (there is only one mold, either male or female).
接触模压成型为开模成型（只有一种模，阳模或者阴模）。

(3) With compression molding, the counter mold will close the mold after the impregnated reinforcements have been placed on the mold.

在模压成型过程中，浸渍的增强体放在模上后，对模（相反的模）会闭合。

(4) Molding by foam injection allows the processing of pieces of fairly large dimensions made of polyurethane foam reinforced with glass fibers.
发泡注射成型允许加工相当大尺寸的玻璃纤维增强的聚氨酯泡沫材料。

Exercises

1. Questions for discussion

(1) What is difference between contact molding and compression molding?
(2) Which molding can process fairly large dimension composite?
(3) Give some examples of compression molding.
(4) How to make the matrix harden?
(5) What is open molding?
(6) What is difference of molding by injection process between thermoset resin and thermoplastic resin.
(7) What is advantage of molding by foam injection?

2. Translate the following into Chinese

polyester resin
hardening
principal process
depression molding
porous fabric
reinforcement
high temperature resistance
accelerator
mold material
contact molding
impregnated with resin
roller to squeeze out the air pockets
compression molding
countermold
ambient temperature
depression molding
bag molding
air bubbles
autoclave
aircraft structures
resin injection molding
polyurethane
centrifugal molding
filament winding
missile tubes
torpillas
containers
reservoirs
propulsion nozzles

(1) The mixture of reinforcement/resin does not really become a composite material until the last phase of the fabrication, that is, when the matrix is hardened.
(2) The process of centrifugal molding is used for the fabrication of tubes. It allows homogeneous distribution of resin with good surface conditions, including the internal surface of the tube.
The process of filament winding can be integrated into a continuous chain of production and can fabricate tubes of long length.
(3) During the solidification process, it passes from the liquid state to the solid state by copolymerization with a monomer that is mixed with the resin.
(4) The layers of fibers impregnated with resin (and accelerator) are placed on the mold.

(5) Compaction is done using a roller to squeeze out the air pockets.

(6) The duration for resin setting varies, depending on the amount of accelerator, from a few minutes to a few hours.

(7) With compression molding, the countermold will close the mold after the impregnated reinforcements have been placed on the mold.

(8) The process is good for average volume production: one can obtain several dozen parts a day (up to 200 with heating).

(9) One sheet of soft plastic is used for sealing (this is adhesively bonded to the perimeter of the mold).

(10) The piece is then compacted due to the action of atmospheric pressure, and the air bubbles are eliminated. Porous fabrics absorb excess resin.

(11) With resin injection molding, the reinforcements (mats, fabrics) are put in place between the mold and counter mold.

(12) Molding by foam injection allows the processing of pieces of fairly large dimensions made of polyurethane foam reinforced with glass fibers.

(13) It allows homogeneous distribution of resin with good surface conditions, including the internal surface of the tube. The length of the tube depends on the length of the mold.

(14) This process is used to fabricate components of high internal pressure, such as reservoirs and propulsion nozzles.

3. Translate the following into English

聚氨酯　　袋压成型

酚醛树脂　　模压成型

浸渍

(1) 成型加工的形式依赖于要成型的部件，部件数量及成本。

(2) 树脂固化的时间依赖于引发剂的量，从几分钟到几小时。

(3) 因为大气压力，树脂被压缩，气泡被除去。

(4) 这种加工方法可用于制造高内部压力的部件，如储水池和推进喷嘴。

Unit 3 Applications of composite materials

Many composites used today are at the leading edge of materials technology, with performance and costs appropriate to ultrademanding applications such as spacecraft. But heterogeneous materials combining the best aspects of dissimilar constituents have been used by nature for millions of years-trees employ cellulosic fibrous components to reinforce a lignin matrix, and arrange the strong fibers in just the correct direction to withstand loads from wind and other environmental sources. Ancient society, imitating nature, used this approach as well: the Book of Exodus speaks of using straw to reinforce mud in brick making, without which the bricks would have almost no strength.

The modern use of fiber-reinforced polymer composites began in the years near World War Ⅱ, with early applications being rocket motor cases and radomes using glass fibers. The Chevrolet Corvette of the 1950's had a fiberglass body (they still do), which wasn't always easy to repair after a collision but did offer an escape from the rusting that afflicts most car bodies. Growth of the market for composites has been very good overall since the 1960's, averaging around 15% per year. By 1979, composites of all types totaled approximately 8 billion pounds, with a value of about 6 billion dollars. Materials selection has always involved a number of compromises for the engineering designer. Of course, the material's properties are extremely important, since the performance of the structure or component to be designed relies in the properties of the material used in its construction. However, properties come at a cost, and the engineer must balance cost factors in making a materials selection. Cost includes not only the base cost of the material itself, but also factors that affect cost indirectly. Advanced materials almost always cost more: they are more expensive on a per-pound basis, they can require extra training to learn new design procedures, the processing can require new tooling and personnel training, and there may be expensive safety and environmental-impact procedures. The improved properties of the material must be able to justify these additional costs, and it is common for the decision to be a difficult one. In the years following their expansion into mainline engineering applications, composites' cost has been dominated by economies of scale: materials produced in relatively small quantities require more expensive handwork and specialty tooling than traditional materials such as steel and aluminum, and are more expensive for this reason alone. This leads to a chicken-and-egg situation in which a new material is not used because it is too expensive, but its costs remain high largely because it is not produced in sufficient quantities. Partly as a result of this, the composites community has had a kind of evangelical outlook, with materials entrepreneurs touting the many advantages composites could bring to engineering design if they were used more aggressively. The drawbacks of composites, most of which result in increased cost of one sort or another, have kept the material from revolutionizing engineering design. At the same time, the advantages of composites are too compelling to ignore, and they have continued to generate good though not superlative growth rates, on the order of 10%-15% per year.

Composites bring many performance advantages to the designer of structural devices, among which we can list.

Composites have high stiffness, strength, and toughness, often comparable with structural metal alloys. Further, they usually provide these properties at substantially less weight than metals: their "specific" strength and modulus per unit weight is near five times that of steel or aluminum. This means the overall structure may be lighter, and in weight-critical devices such as airplanes or spacecraft this weight savings might be a compelling advantage.

Composites can be made anisotropic, i.e., have different properties in different directions, and this can be used to design a more efficient structure. In many structures the stresses are also different in different directions; For instance in closed-end pressure vessels – such as a rocket motor case –

the circumferential stresses are twice the axial stresses. Using composites, such a vessel can be made twice as strong in the circumferential direction as in the axial.

Many structures experience fatigue loading, in which the internal stresses vary with time. Axles on rolling stock are examples; Here the stresses vary sinusoidally from tension to compression as the axle turns. These fatigue stresses can eventually lead to failure, even when the maximum stress is much less than the failure strength of the material as measured in a static tension test. Composites have excellent fatigue resistance in comparison with metal alloys, and often show evidence of accumulating fatigue damage, so that the damage can be detected and the part replaced before a catastrophic failure occurs.

Materials can exhibit damping, in which a certain fraction of the mechanical strain energy deposited in the material by a loading cycle is dissipated as heat. This can be advantageous, for instance in controlling mechanically-induced vibrations. Composites generally offer relatively high levels of damping, and furthermore the damping can often be tailored to desired levels by suitable formulation and processing.

Composites can be excellent in applications involving sliding friction, with tribological ("wear") properties approaching those of lubricated steel.

Composites do not rust as do many ferrous alloys, and resistance to this common form of environmental degradation may offer better life-cycle cost even if the original structure is initially more costly.

Many structural parts are assembled from a number of subassemblies, and the assembly process adds cost and complexity to the design. Composites offer a lot of flexibility in processing and property control, and this often leads to possibilities for part reduction and simpler manufacture. Of course, composites are not perfect for all applications, and the designer needs to be aware of their drawbacks as well as their advantages. Among these cautionary notes we can list:

Not all applications are weight-critical. If weight-adjusted properties not relevant, steel and other traditional materials may work fine at lower cost.

Anisotropy and other "special" features are advantageous in that they provide a great deal of design flexibility, but the flip side of this coin is that they also complicate the design. The well-known tools of stress analysis used in isotropic linear elastic design must be extended to include anisotropy, for instance, and not all designers are comfortable with these more advanced tools.

Even after several years of touting composites as the "material of the future," economies of scale are still not well developed. As a result, composites are almost always more expensive-often much more expensive-than traditional materials, so the designer must look to composites' various advantages to offset the extra cost. During the energy-crisis period of the 1970's, automobile manufacturers were so anxious to reduce vehicle weight that they were willing to pay a premium for composites and their weight advantages. But as worry about energy efficiency diminished, the industry gradually returned to a strict lowest-cost approach in selecting materials. Hence the market for composites in automobiles returned to a more modest rate of growth.

Although composites have been used extensively in demanding structural applications for a half-century, the long-term durability of these materials is much less certain than that of steel or other traditional structural materials. The well-publicized separation of the tail fin of an American Airlines A300-600 Airbus after takeoff from JFK airport on November 12, 2001 is a case in point. It is not clear that this accident was due to failure of the tail's graphite-epoxy material, but NASA is looking very hard at this possibility. Certainly there have been media reports expressing concern about the material, and this point up the uncertainty designers must consider in employing composites.

New words and expressions

cellulosic *adj.* 纤维素的
fibrous *adj.* 纤维的，纤维构成的，纤维状的
the Book of Exodus 出埃及记
rocket motor case 火箭发动机壳体
radome *n.* 雷达天线罩
Chevrolet Corvette 雪佛兰考维特
rust *n.* 生锈，*v.* 生锈
evangelical *n.*新教徒，福音派教徒；*adj.* 福音的，新教的
entrepreneur *n.* 企业家
superlattic *n.* 超晶格，超点阵
circumferential *adj.* 周向，圆周
fatigue *n.* 疲劳
sinusoidally *adv.* 正弦型
catastrophic *adj.* 悲惨的，灾难的，爆炸性的
damping 阻尼，衰减，内耗
tribological *adj.* 摩擦学的
graphite *n.* 石墨，石墨纤维

Notes

(1) The Chevrolet Corvette of the 1950's had a fiberglass body (they still do), which wasn't always easy to repair after a collision but did offer an escape from the rusting that afflicts most car bodies.
20 世纪 50 年代的雪佛兰考维特为玻璃纤维的外壳（现在仍是），碰撞后不容易修复，但是能避免大多数车体所遇到的生锈问题。

(2) This leads to a chicken-and-egg situation in which a new material is not used because it is too expensive, but its costs remain high largely because it is not produced in sufficient quantities.
这导致谁先谁后的状况，即一种新材料不能应用因为它价格太昂贵，它的价格很高主要因为没有大批量生产。

(3) At the same time, the advantages of composites are too compelling to ignore, and they have continued to generate good though not superlative growth rates, on the order of 10%-15% per year.
同时，复合材料的优势如此引人注目而不能忽视，它们持续产生较好的尽管不是最快的增长速度，大约每年 10%~15%。

(4) Composites can be made *anisotropic,* i.e., have different properties in different directions, and this can be used to design a more efficient structure.
复合材料能够制作成各向异性，也就是在不同的方向有不同的性质，这能被用来设计一

种更为有效的结构。

(5) Axles on rolling stock are examples; Here the stresses vary sinusoidally from tension to compression as the axle turns.
车辆的车轴为例，当车轴运转时存在从拉伸到压缩正弦变化的应力。

(6) The well-publicized separation of the tail fin of an American Airlines A300-600 Airbus after takeoff from JFK airport on November 12, 2001 is a case in point.
广泛报道的在 2001 年 11 月 12 号 JFK 机场起飞的美国航空公司 A300-600 空中客车尾翼分离事故是一个恰当的例子。

Exercises

1. Questions for discussion

(1) What are the early applications of fiber-reinforced polymer composites?
(2) What are the shortages of anisotropy of polymer composites?
(3) Composites are not perfect for all applications, please offer an example.
(4) Composites bring many performance advantages to the designer of structural devices, please elaborate these advantages.
(5) Composite materials have been used by nature for a long time, please list some examples.
(6) What is the reason that advanced materials almost always cost more?

2. Translate the following into Chinese

Heterogeneous	static tension test	heterogeneous materials
lignin matrix	failure strength	radomes
rocket motor cases	loading cycle	sinusoidally
entrepreneurs	per-pound	fatigue resistance
circumferential stresses	ultrademanding applications	damping

(1) Even after several years of touting composites as the "material of the future," economies of scale are still not well developed.
(2) Many structural parts are assembled from a number of subassemblies, and the assembly process adds cost and complexity to the design.
(3) Of course, the material's properties are extremely important, since the performance of the structure or component to be designed relies in the properties of the material used in its construction.
(4) The drawbacks of composites, most of which result in increased costs of one sort or another, have kept the material from revolutionizing engineering design.
(5) However, properties come at a cost, and the engineer must balance cost factors in making a materials selection.
(6) Composites have high stiffness, strength, and toughness, often comparable with structural metal alloys.
(7) Composites can be made anisotropic, i.e., have different properties in different directions, and this can be used to design a more efficient structure.

(8) Composites have excellent fatigue resistance in comparison with metal alloys, and often show evidence of accumulating fatigue damage, so that the damage can be detected and the part replaced before a catastrophic failure occurs.
(9) Materials can exhibit damping, in which a certain fraction of the mechanical strain energy deposited in the material by a loading cycle is dissipated as heat.
(10) Composites do not rust as do many ferrous alloys, and resistance to this common form of environmental degradation may offer better life-cycle cost even if the original structure is initially more costly.
(11) Composites offer a lot of flexibility in processing and property control, and this often leads to possibilities for part reduction and simpler manufacture.
(12) As a result, composites are almost always more expensive – often much more expensive – than traditional materials, so the designer must look to composites' various advantages to offset the extra cost.
(13) Although composites have been used extensively in demanding structural applications for a half-century, the long-term durability of these materials is much less certain than that of steel or other traditional structural materials.

3. Translate the following into English

拉伸	应力分析	各向同性
阻尼	使用周期	
各向异性	合金	

(1) 材料改进后的性能，必须能够弥补这些额外的成本，这通常是比较难以决定。
(2) 相比于合金，复合材料具有优良的耐疲劳性，通常可表现出累积疲劳损伤的证据，这样损伤能被检测，损坏部件可以在灾难性故障发生前替换。
(3) 复合材料不像铁合金那样容易生锈，对生锈这种常见环境降解形式的抵抗能够提供更好的寿命周期成本，即使最初的结构成本昂贵。

4. Scenario simulation

Supposing you are college senior student majoring in composite materials. The department of composite material which you study in will hold a freshman orientation to the freshmen of this department. Please introduce fiber reinforced polymers to the freshmen of composite material by PowerPoint. The content should include the basic concept, process method, and apply of fiber reinforced polymers.

在线习题

拓展阅读

Chapter 6 Metal matrix composite

扫码听音频

Unit 1 Introduction

Metal matrix composites (MMCs), like all composites, consist of at least two chemically and physically distinct phases, suitably distributed to provide properties not obtainable with either of the individual phases. Generally, there are two phases, e.g., a fibrous or particulate phase, distributed in a metallic matrix. Examples include continuous Al_2O_3 fiber reinforced Al matrix composites used in power transmission lines; Nb-Ti filaments in a copper matrix for superconducting magnets; tungsten carbide (WC)/cobalt (Co) particulate composites used as cutting tool and oil drilling inserts; and SiC particle reinforced Al matrix composites used in aerospace, automotive, and thermal management applications.

A legitimate question that the reader might ask is: Why metal matrix composites? The answer to this question can be subdivided into two parts: (a) advantages with respect to unreinforced metals and (b) advantages with respect to other composites such as polymer matrix composites (PMCs). With respect to metals, MMCs offer the following advantages: major weight savings due to higher strength-to-weight ratio; exceptional dimensional stability (compare, for example, SiC/Al to Al); higher elevated temperature stability, i.e., creep resistance; significantly improved cyclic fatigue characteristics.With respect to PMCs, MMCs offer these distinct advantages: higher strength and stiffness; higher service temperatures; higher electrical conductivity (grounding, space charging); higher thermal conductivity; better transverse properties; improved joining characteristics; radiation survivability (laser, UV, nuclear, etc.); little or no contamination (no out-gassing or moisture absorption problems).

(1) Types of MMCs All metal matrix composites have a metal or a metallic alloy as the matrix. The reinforcement can be metallic or ceramic. In some unusual cases, the composite may consist of a metallic alloy "reinforced" by a fiber reinforced polymer matrix composite (e.g., a sheet of glass fiber reinforced epoxy or aramid fiber reinforced epoxy).

In general, there are three kinds of metal matrix composites (MMCs): particle reinforced MMCs; short fiber or whisker reinforced MMCs; continuous fiber or sheet reinforced MMCs.

Fig. 6.1 shows, schematically, the three major types of metal matrix composites: Continuous fiber reinforced, short fiber or whisker reinforced, particle reinforced, and laminated or layered composites. The reader can easily visualize that the continuous fiber reinforced composites will be the most anisotropic of all. Tab. 6.1 provides examples of some important reinforcements used in metal matrix composites as well as their aspect ratios (length / diameter) and diameters.

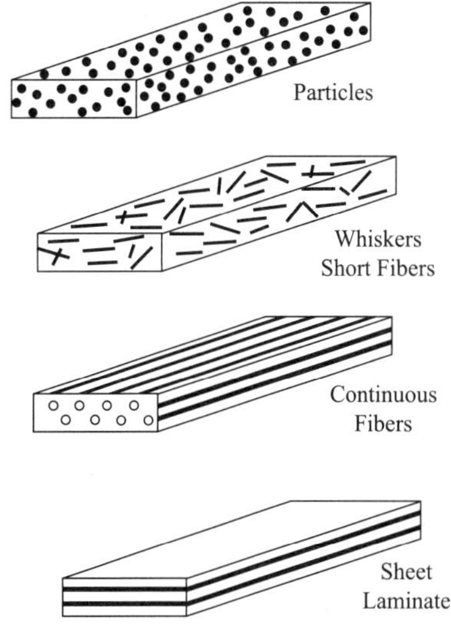

Fig. 6.1 Different types of metal matrix composites

Particle or discontinuously reinforced MMCs (the term discontinuously reinforced MMCs is commonly used to indicate metal matrix composites having reinforcements in the form of short fibers, whiskers, or particles) have assumed special importance for the following reasons:

① Particle reinforced composites are inexpensive ***vis-a-vis*** continuous fiber reinforced composites. Cost is an important and essential parameter, particularly in applications where large volumes are required (e.g., automotive applications).

Tab. 6.1 Typical reinforcements used in metal matrix composites

Type	Aspect ratio	Diameter/μm	Examples
Particle	1-4	1-25	SiC, Al_2O_3, BN, B_4C, WC
Short fiber or whisker	10-10000	1-5	C, SiC, Al_2O_3, Al_2O_3+SiO_2
Continuous fiber	>1000	3-150	SiC, Al_2O_3, C, B, W, Nb-Ti, Nb_3Sn

② Conventional metallurgical processing techniques such as casting or powder metallurgy, followed by conventional secondary processing by rolling, forging, and extrusion can be used.

③ Higher use temperatures than the unreinforced metal.

④ Enhanced modulus and strength.

⑤ Increased thermal stability.

⑥ Better wear resistance.

⑦ Relatively isotropic properties compared to fiber reinforced composites.

Within the broad category of discontinuously reinforced composites, metal matrix composites made by liquid metal casting are somewhat cheaper to produce than powder metallurgy composites. There are two types of cast metal matrix composites.

⑧ Cast composites having local reinforcement.

⑨ Cast composites in the form of a billet having uniform reinforcement with a wrought alloy matrix. Such composite billets are forged and/or extruded, followed by rolling or other forming operations.

(2) Characteristics of MMCs One of the driving forces for metal matrix composites is, of course, enhanced stiffness and strength. There are other characteristics which may be equally valuable. As examples, we can cite the ability to control thermal expansion in applications involving electronic packaging. By adding ceramic reinforcements, one can generally reduce the coefficient of linear thermal expansion of the composite. Electrical and thermal conductivity characteristics may be important in some applications.

Clearly, superconductors require superconducting characteristics. The metallic matrix provides a high thermal conductivity medium in case of an accidental quench, in addition to holding the tiny superconducting filaments together. Other important characteristics that may be of immense value include wear resistance (e.g., WC/Co composites used in cutting tools or oil drilling inserts and SiCp/Al rotor in brakes). Thus, although one commonly uses the term reinforcement by particle or fibers in the context of metal matrix composites, it worth pointing out that strength enhancement may not be the most important characteristic in many applications. In the chapters that follow we explore these and other unique and important attributes of metal matrix composites.

(3) Composition of MMCs MMCs are made by dispersing a reinforcing material into a metal matrix. The reinforcement surface can be coated to prevent a chemical reaction with the matrix. For example, carbon fibers are commonly used in aluminium matrix to synthesize composites showing low density and high strength. However, carbon reacts with aluminum to generate a brittle and water-soluble compound Al_4C_3 on the surface of the fiber. To prevent this reaction, the carbon fibers are coated with nickel or titanium boride.

① **Matrix** The matrix is the monolithic material into which the reinforcement is embedded, and is completely continuous. This means that there is a path through the matrix to any point in the material, unlike two materials sandwiched together. In structural applications, the matrix is usually a lighter metal such as aluminum, magnesium, or titanium, and provides a compliant support for the reinforcement. In high temperature applications, cobalt and cobalt-nickel alloy matrices are common.

② **Reinforcement** The reinforcement material is embedded into the matrix. The reinforcement does not always serve a purely structural task (reinforcing the compound), but is also used to change physical properties such as wear resistance, friction coefficient, or thermal conductivity. The reinforcement can be either continuous, or discontinuous. Discontinuous MMCs can be isotropic, and can be worked with standard metalworking techniques, such as extrusion, forging or rolling. In addition, they may be machined using conventional techniques, but commonly would need the use of polycrystalline diamond tooling (PCD).

Continuous reinforcement uses monofilament wires or fibers such as carbon fiber or silicon carbide. Because the fibers are embedded into the matrix in a certain direction, the result is an anisotropic structure in which the alignment of the material affects its strength. One of the first MMCs used boron filament as reinforcement. Discontinuous reinforcement "uses whiskers", short fibers, or particles. The most common reinforcing materials in this category are alumina and silicon carbide.

(4) Manufacturing and forming methods of MMCs MMC manufacturing can be broken into three types: solid, liquid, and vapor.

① **Solid state methods**

a. Powder blending and consolidation (powder metallurgy): Powdered metal and discontinuous reinforcement are mixed and then bonded through a process of compaction, degassing, and thermo-mechanical treatment [possibly via hot isostatic pressing (HIP) or extrusion].

b. Foil diffusion bonding: Layers of metal foil are sandwiched with long fibers, and then pressed through to form a matrix.

② **Liquid state methods**

a. Electroplating/Electroforming: A solution containing metal ions loaded with reinforcing particles is co-deposited forming a composite material.

b. Stir casting: Discontinuous reinforcement is stirred into molten metal, which is allowed to solidify.

c. Squeeze casting: Molten metal is injected into a form with fibers preplaced inside it.

d. Spray deposition: Molten metal is sprayed onto a continuous fiber substrate.

e. Reactive processing: A chemical reaction occurs, with one of the reactants forming the matrix and the other the reinforcement.

③ **Vapor deposition**

a. Physical vapor deposition: The fiber is passed through a thick cloud of vaporized metal, coating it.

b. In situ fabrication technique.

c. Controlled unidirectional solidification of a eutectic alloy can result in a two-phase microstructure with one of the phases, present in lamellar or fiber form, distributed in the matrix.

New words and expressions

composite *n.* 复合材料
metal matrix *n.* 金属基
filaments *n.* 丝状物
dimensional *adj.* 空间的，尺寸的
transmission *n.* 播送，传送
carbide *n.* 碳化物
legitimate *adj.* 合理的
transverse *adj.* 横向的
laminated *adj.* 层压的，薄板状的
strength-to-weight ratio *n.* 比强度
creep resistance *n.* 抗蠕变性
forging *n.* 锻造

extrusion *n.* 挤出，挤压

quench *n.* 淬火，冷浸

whisker reinforced *adj.* 晶须增强的

electroplate *vt.* 电镀；*n.*电镀物品

vapor deposition *n.* 蒸镀，气相沉积

Notes

(1) Metal matrix composites (MMCs), like all composites, consist of at least two chemically and physically distinct phases, suitably distributed to provide properties not obtainable with either of the individual phases.
金属基复合材料（MMCs），就像其他所有的复合材料一样，至少包括化学和物理性能不同的两个相，并通过合理分布来提供一些单相无法获得的性能。

(2) For example, carbon fibers are commonly used in aluminum matrix to synthesize composites showing low density and high strength.
例如，碳纤维常被用于铝合金基体中以合成低密度，高强度的复合材料。

(3) Conventional metallurgical processing techniques such as casting or powder metallurgy, followed by conventional secondary processing by rolling, forging, and extrusion can be used.
传统的冶金工艺技术，比如说铸造或者粉末冶金，紧随其后的是传统二次加工，如轧制，锻造和挤出，都可以使用。

(4) The metallic matrix provides a high thermal conductivity medium in case of an accidental quench, in addition to holding the tiny superconducting filaments together.
除了可将微小的超导细丝结合在一起外，金属基体还提供了一个高的导热介质以防止意外终止的情况。

(5) Controlled unidirectional solidification of a eutectic alloy can result in a two-phase microstructure with one of the phases, present in lamellar or fiber form, distributed in the matrix.
共晶合金的可控定向凝固能使其中一相中含有两相微结构，即以薄片或纤维的形式分布在基体中。

Exercises

1. Question for discussion

(1) What is metal matrix composite? Can you give some examples of the metal matrix composites?
(2) Which two phases do the metal matrix composites have?
(3) Which industries need aluminum matrix composite material?
(4) What is the role of the reinforcement, in addition to serving a structural task?
(5) What is metal matrix composites used for by adding ceramic reinforcement?

2. Translate the following into Chinese

coefficient of linear thermal expansion　liquid metal casting
wear resistance　friction coefficient
polycrystalline diamond tooling　thermo-mechanical treatment

(1) Within the broad category of discontinuously reinforced composites, metal matrix composites made by liquid metal casting are somewhat cheaper to produce than powder metallurgy composites.
(2) By adding ceramic reinforcements, one can generally reduce the coefficient of linear thermal expansion of the composite.
(3) Because the fibers are embedded into the matrix in a certain direction, the result is an anisotropic structure in which the alignment of the material affects its strength.

3. Translate the following into English

热稳定性	电子封装	热等静压
锻造合金	贴膜扩散接合	喷射沉积

(1) 基体材料是金属基复合材料的主要组成部分，起着固结增强物、传递和承受各种载荷(力、热、电)的作用。
(2) 增强体是金属基复合材料的重要组成部分，它起着提高金属基体的强度、模量、耐热、耐磨等性能的作用。
(3) 界面结构对金属基复合材料性能具有很大影响，界面结合力的大小也对金属基复合材料产生极大的影响。
(4) 固相法就是将金属粉末或金属箔与增强物（纤维、晶须、颗粒等）按设计要求以一定的含量、分布、方向混合排布在一起，再经加热、加压，将金属基体与增强物复合黏结在一起形成金属基复合材料。

4. Scenario simulation

Supposing you are an intern in a composites company, a customer hopes to discuss business about metal matrix composites with you. Please write a report on the introduction to metal matrix composites for the customer in no more than 200 words.

Unit 2 Processing

Metal matrix composites can be made by liquid, solid, or gaseous state processes. In this section we describe some important processing techniques for fabricating MMCs.

(1) Liquid state processing Metal matrix composites can be processed by incorporating or combining a liquid metal matrix with the reinforcement. There are several advantages to using a liquid phase route in processing. These include near net-shape (when compared to solid state processes like extrusion or diffusion bonding), faster rate of processing, and the relatively low temperatures associated with melting most light metals, such as Al and Mg. The most common liquid phase processing techniques can be subdivided into four major categories.

① **Casting or liquid infiltration** This involves infiltration of a fibrous or particulate preform by a liquid metal. In the case of direct introduction of short fibers or particles into a liquid mixture, consisting of liquid metal and ceramic particles or short fibers, is often stirred to obtain a homogeneous distribution of particles. In centrifugal casting, a gradient in reinforcement particle loading is obtained. This can be quite advantageous from a machining or performance perspective.

Example: Conventional casting

Casting of MMCs can typically be accomplished with conventional equipment used to cast metallic alloys (Surappa and Rohatgi, 1981). It is typically used with particulate reinforcement because of the difficulty in casting fibrous performs without pressure. The particles and matrix mixture are cast into ingots and a secondary mechanical process, such as extrusion or rolling, is applied to the composite, see Fig. 6.2.

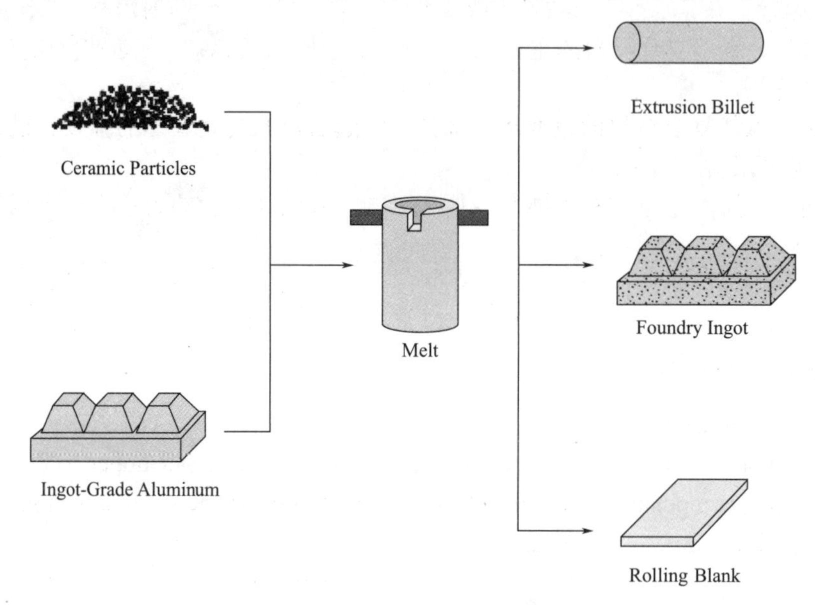

Fig. 6.2 Conventional casting route for processing particle reinforced MMCs

② **Squeeze casting or pressure infiltration** This method encompasses pressure-assisted liquid infiltration of a fibrous or particulate preform. This process is particularly suited for complex shaped components, selective or localized reinforcement, and where production speed is critical.

Example: Squeeze casting

Squeeze casting or pressure infiltration involves forcing the liquid metal matrix into a short fiber or particulate preform (Mortensen et al., 1988; Masur et al., 1989; Cook and Werner, 1991). The main advantages of this method over conventional casting are the shorter processing times (which is of particular interest for production of materials in high volumes), ability to fabricate relatively complex shapes, minimal residual porosity or shrinkage cavities due to the applied pressure, and minimization of interfacial reaction products between reinforcement and matrix (due to the shorter processing times). Before infiltration takes place, the reinforcement preform must be prepared.

In order to obtain infiltration of the preform, the molten metal must have a relatively low viscosity and good wettability of the reinforcement. A schematic of the liquid infiltration process is shown in Fig. 6.3. The reinforcement preform is placed in a mold, and the liquid is poured into a preheated die located on the bed of a hydraulic press. Infiltration takes place by mechanical force or by using a pressurized inert gas. Applied pressures on the order of 70-100 MPa are typically used. Having the preform temperature lower than that of the matrix liquidus temperature is highly desirable in order to minimize interfacial reaction and to obtain a fine matrix grain size.

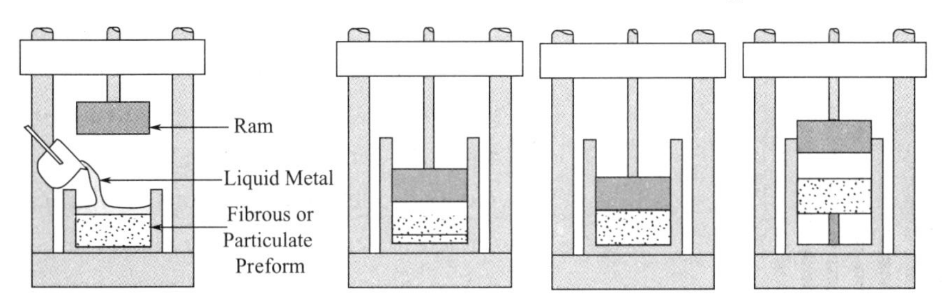

Fig. 6.3 Schematic of squeeze casting process

a. Spray co-deposition In this process the liquid metal is atomized or sprayed while a particle injector introduces ceramic particles in the spray stream to produce a granulated mixture of composite particles. The composite particles are then consolidated using another suitable technique, such as hot-pressing, extrusion, forging, etc.

Spray deposition has been used for some time to fabricate metallic alloys in powder form (Lavernia et al.,1992). The metal or alloy is melted and the liquid stream is atomized with water or an inert gas. Rapid solidification of the liquid takes place, resulting in a fine solid powder. This technique has been modified, by injecting reinforcement particles or co-depositing the particles with the matrix alloy, Fig. 6.4 (Lloyd, 1997). The advantage of this technique is the high rate of production, which can approach 6-10 kg/min, and the very fast solidification rate, which minimizes any reaction between particle and matrix.

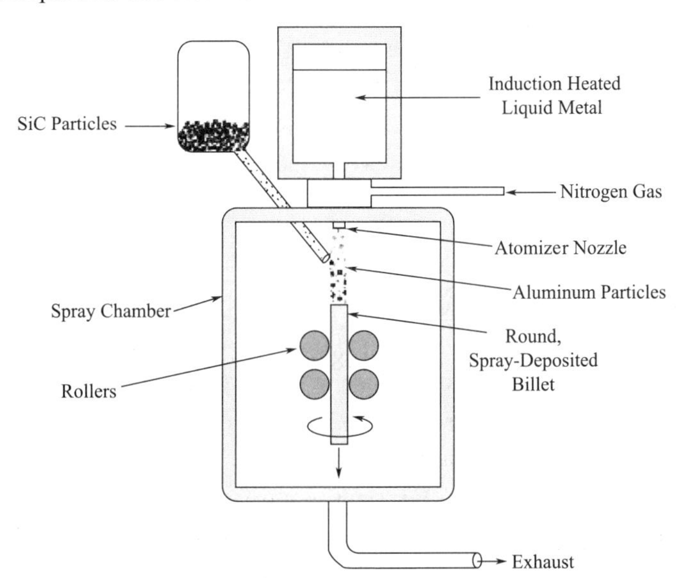

Fig. 6.4 Spray co-deposition of SiC particles and Al liquid droplets, to form composite particles

b. In situ processes In this case, the reinforcement phase is formed in situ either by reaction during synthesis or by controlled solidification of a eutectic alloy.

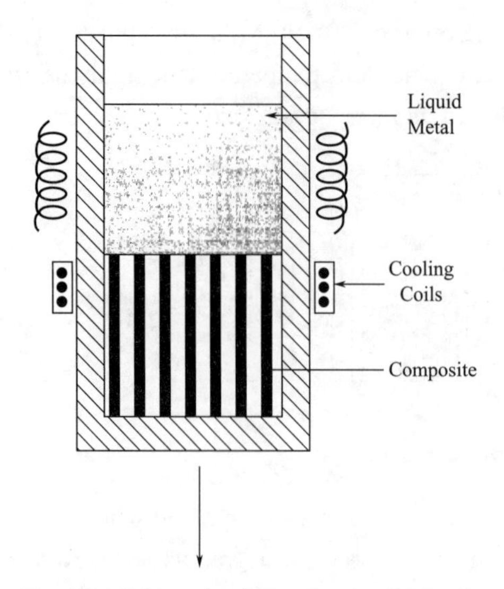

Fig. 6.5 Schematic of directional solidification process to obtain in situ composites

In situ processes fall into two major categories: Reactive and non-reactive processes. In the reactive processes two components are allowed to react exothermically to form the reinforcement phase. Non-reactive in situ processes take advantage of two-phase systems, such as eutectic or monotectic alloys, to form the fiber and matrix in situ (McLean, 1983). Controlled directional solidification is conducted to separate the two phases, as shown in Fig. 6.5. A precast and homogenized material is melted in a graphite crucible, and contained in a quartz tube in vacuum or inert gas atmosphere. Heating is typically conducted by induction and the thermal gradient is obtained by chilling the crucible. Electron beam heating may also be used, particular when using reactive metals such as titanium.

(2) Solid state processing The main drawback associated with liquid phase techniques is the difficulty in controlling reinforcement distribution and obtaining a uniform matrix microstructure (Michaud, 1993). Furthermore, adverse interfacial reactions between the matrix and the reinforcement are likely to occur at the high temperatures involved in liquid processing. These reactions can have an adverse effect on the mechanical properties of the composite (Sahoo and Koczak, 1991; Chawla, 1997). The most common solid phase processes are based on powder metallurgy techniques (Ghosh, 1993). These typically involve discontinuous reinforcements, due to the ease of mixing and blending, and the effectiveness of densification. The ceramic and metal powders are mixed, isostatically cold compacted, and hot-pressed to full density. The fully-dense compact then typically undergoes a secondary operation such as extrusion or forging (Lloyd, 1997). Novel low-cost approaches, such as sinter-forging, have aimed at eliminating the hot pressing step, with promising results (Chawla et al., 2003).

① **Powder metallurgy processing** Powder processing involves cold pressing and sintering, or hot pressing to fabricate primarily particle- or whisker-reinforced MMCs (Hunt,1994). The matrix and the reinforcement powders are blended to produce a homogeneous distribution. The blending stage is followed by cold pressing to produce what is called a green body, which is about 80% dense and can be easily handled, Fig.6.6. The cold pressed green body is canned in a container, sealed, and degassed to remove any absorbed moisture from the particle surfaces. One of the problems with bonding metallic powder particles, such as A1 particles, to ceramic particles, such as SiC, or to other A1 particles is the oxide "skin" that is invariably present on the Al particle surface (Kim et al.,1985; Anderson and Foley, 2001). Degassing and hot pressing in an inert atmosphere contributes to the removal of Al hydrides present on the particle surface, making the oxide skin more brittle and, thus, more easily sheared (Estrada et al., 1991; Kowalski et al., 1992). The material is hot pressed, uniaxially or isostatically, to produce a fully dense composite and extruded. The rigid particles or fibers do not deform, causing the matrix to be deformed significantly.

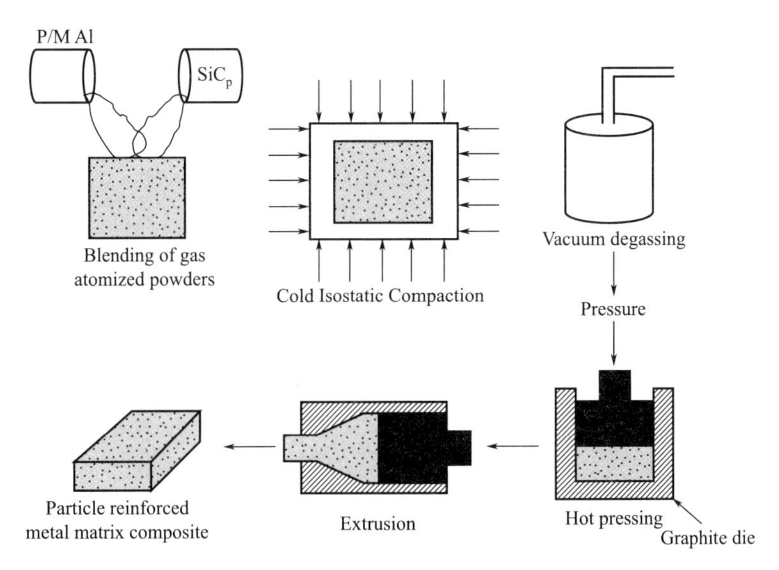

Fig. 6.6 Powder processing, hot pressing, and extrusion process for fabricating particulate or short fiber reinforced MMCs

② **Extrusion** Extrusion processing has been used extensively as a means of secondary deformation processing of MMCs (Ghosh, 1993; Hunt, 1994; Lloyd, 1997). It is particularly advantageous because the combination of pressure and temperature results in shear between Al/Al particles and Al/SiC particles, which contributes to fracture of the oxide skin on the Al particles, and the bonding between particle and matrix is enhanced. Because of the large strains associated with this process, however, extrusion has been used primarily to consolidate composites with discontinuous reinforcement, in order to minimize reinforcement fracture. Even in discontinuously reinforced materials, fracture of short fibers or particles often takes place, which can be detrimental to the properties of the composite.

③ **Forging** Forging is another common secondary deformation processing technique used to manufacture metal matrix composites. Once again, this technique is largely restricted to composites with discontinuous reinforcement. In conventional forging, a hot-pressed or extruded product is forged to near-net shape (Helinski et al., 1994).

④ **Pressing and sintering** A relatively inexpensive and simple technique involves pressing and sintering of powders. These composite systems are typically sintered in a temperature range to obtain some degree of liquid phase. The liquid phase flows through the pores in the compact resulting in densification of the composite (unless interfacial reaction takes place). Special mention should be made of WC/Co composites, commonly known as cemented carbides. They are really nothing but very high volume fraction of WC particles distributed in a soft cobalt matrix. These composites are used extensively in machining and rock and oil drilling operations.

⑤ **Roll bonding and co-extrusion** Roll bonding is a common technique used to produce a laminated composite consisting of different metals in layered form (Chawla and Godefroid, 1984). Such composites are called sheet laminated metal-matrix composites. Roll bonding and hot pressing have also been used to make laminates of Al sheets and discontinuously reinforced MMCs (Hunt et al., 1991; Manoharan et al., 1990). Fig. 6.7 shows the roll bonding process of making a laminated MMC.

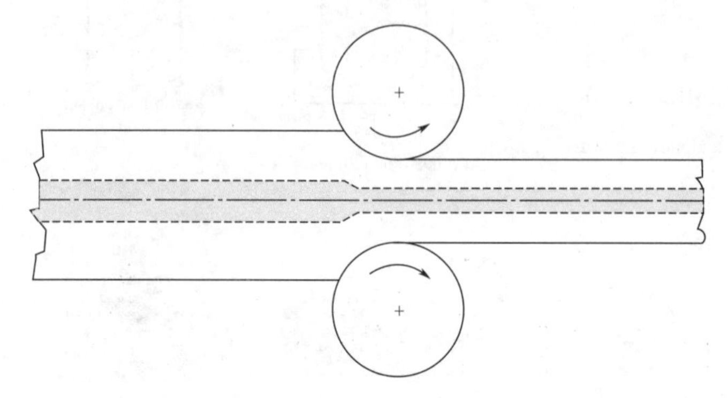

Fig. 6.7 Roll bonding process of making a laminated MMC where a metallurgical bond is produced between the layers

⑥ **Diffusion bonding** Diffusion bonding is a common solid-state processing technique for joining similar or dissimilar metals. Interdiffusion of atoms, at an elevated temperature, from clean metal surfaces in contact with each other leads to bonding (Partridge and Ward-Close, 1993; Guo and Derby, 1995). The principal advantages of this technique are the ability to process a wide variety of matrix metals and control of fiber orientation and volume fraction. Among the disadvantages are long processing times, high processing temperatures and pressures (which makes the process expensive), and limitation on complexity of shapes that can be produced.

⑦ **Explosive shock consolidation** A fairly novel, high strain rate, rapid solidification technique is explosive shock consolidation (Thadhani, 1988; and Thadhani et al., 1991). In this technique, dynamic compaction and/or synthesis of powders can be achieved by means of shock waves generated by either explosives in contact with the powder or high velocity impact from projectiles. This process is particularly attractive for consolidating hard materials such as ceramics, or composites.

(3) Gaseous state processing Plasma spraying is the primary form of gaseous state processing. The main application of plasma spraying was described above, to form matrix-coated fibers, which are subsequently hot-pressed to form the final product. In addition, laminated composites, particularly on the nanometer scale have been processed by physical vapor deposition process (PVD). PVD processes (specifically, sputter deposition-based processes) offer an extremely wide range of possibilities for fabricating nanolaminate microstructures with tailored chemistry, structure, and thickness of the individual layers and interfaces. Additional important PVD processing parameters include reactive deposition (Ji et al., 2001), plasma-assisted deposition (O' Keefe and Rigsbee, 1994), and substrate heating (Misra et al., 1998).

Some metal/metal layered systems, such as Ni/Cu and Ni/Ti, have been processed at the nanoscale by sputter deposition with great success (Misra et al., 1998; Misra and Nastasi, 1999). An example of a nanoscale Cu-Ni multilayer with a bilayer period of 5 nm was prepared (Misra et al.,

1998). Note the well-defined layered structure. The corresponding selected area diffraction pattern (inset) shows a (001) growth direction and cube-oncube orientation relationship between FCC Cu and FCC Ni. A challenge in the synthesis of nanolaminates via the PVD approach is the control of intrinsic residual stresses. Some control over residual stresses has been achieved by energetic particle bombardment, either in situ or post-deposition using an ion source. In the case of magnetron sputtering, a negative substrate bias may be sufficient to change the residual stress from tensile to compressive (Misra and Nastasi, 1999). A film deposited with low bombardment energy yields a tensile residual stress and a microstructure with nanoscale columnar porositylcracking for a sputtered 150 nm thick Cr film. The same material sputtered with a negative bias, on the other hand, yields a nanocrystalline film with an equiaxed grain structure, near-zero residual stress, and, thus, no intergranular porosity.

New words and expressions

subdivide *vt.* 细分
infiltration *n.* 渗透；渗滤
homogeneous *adj.* 同性质的，同类的，均匀的
centrifugal *n.* 离心
gradient *n.* 梯度，陡度；*adj.* 倾斜的
ingots *n.* 钢锭，铸块
viscosity *n.* 黏性；黏质
wettability *n.* 润湿性
mold *n.* 模子；*vt.* 浇铸
hydraulic press *n.* 水压机
granulated *adj.* 颗粒状的
components *n.* 部分，组件
exothermic *adj.* 放出热量的
compacted *adj.* 压实的，压紧的
eutectic *adj.* 共熔的；*n.* 共晶
monotectic *n.* 偏共晶，偏晶体
uniaxially *adv.* 单向地
cracking *n.* 裂纹，裂缝

Notes

(1) The particles and matrix mixture are cast into ingots and a secondary mechanical process, such as extrusion or rolling, is applied to the composite，see Fig. 6.2.
如图 6.2 所示，颗粒和基质混合经浇铸成锭和二次机械加工，如挤压或轧制，被应用于复合材料的加工中。

(2) This process is particularly suited for complex shaped components, selective or localized reinforcement, and where production speed is critical.
这种方法特别适合于形状复杂的零件，选择性或局部加固，生产速度在这一步非常重要。

(3) In this process the liquid metal is atomized or sprayed while a particle injector introduces ceramic particles in the spray stream to produce a granulated mixture of composite particles.
在这一工艺中，液态金属雾化或喷涂，同时喷射流中粒子喷射器将陶瓷颗粒制成颗粒状混合物的复合粒子。

(4) The most common solid phase processes are based on powder metallurgy techniques (Ghosh, 1993). These typically involve discontinuous reinforcements, due to the ease of mixing and blending, and the effectiveness of densification.
最常见的固相法是基于粉末冶金技术(Ghosh, 1993)。由于易于混合和搅拌，固相法最大的

特点就是包含不连续的增强体，而且制备的复合材料致密度高。

(5) Degassing and hot pressing in an inert atmosphere contributes to the removal of Al hydrides present on the particle surface, making the oxide skin more brittle and, thus, more easily sheared.
在惰性气氛脱气和热压有助于去除颗粒表面的铝氢化物，使氧化皮更脆，从而更容易被脱去。

Exercises

1. Question for discussion

(1) Which processes can be used to fabricate metal matrix composite?
(2) What kinds of main liquid phase processing are there?
(3) Please simply describe casting or liquid infiltration and give some examples.
(4) Comparing with liquid phase techniques, what is the advantage of solid phase techniques?

2. Translate the following into Chinese

light metals	centrifugal casting
interfacial reaction	spray co-deposition
electron beam	plasma-assisted deposition

(1) The advantage of this technique is the high rate of production, which can approach 6~10 kg/min, and the very fast solidification rate, which minimizes any reaction between particle and matrix.
(2) It is particularly advantageous because the combination of pressure and temperature results in shear between Al/Al particles and Al/SiC particles, which contributes to fracture of the oxide skin on the Al particles, and the bonding between particle and matrix is enhanced.
(3) Among the disadvantages are long processing times, high processing temperatures and pressures (which makes the process expensive), and limitation on complexity of shapes that can be produced.

3. Translate the following into English

衬底	互扩散	二次成型
等离子喷涂	冷压	铝片层压板

(1) 如果将含有液态金属、陶瓷粒子或短纤维的直接引入液态混合物，常常通过充分搅拌来实现粒子的均匀分布。
(2) 为了实现粗产品的相互渗透，熔融的金属需有相对较低的黏度，还要保持与增强体之间良好的润湿性。
(3) 由于挤出过程中大的应力，挤出工艺主要被用来与不连续的增强体一起合成复合材料，从而最大限度减少增强体的断裂。

Unit 3 Applications

Metal matrix composites are used in a myriad of applications. The high strength-to-weight ratio, enhanced mechanical and thermal properties over conventional materials, and tailorability of properties make them very attractive in a variety of applications. Increasingly MMCs have been used in several areas including (Evans et al, 2003): Aerospace; Transportation (automotive and railway); Electronics and thermal management; Filamentary superconducting magnets; Power conduction; Recretational Products and Sporting Goods; Wear-resistant materials.

In this section we review some important applications of MMCs, and point out the advantages of using MMCs.

(1) Aerospace In aerospace applications, low density, tailored thermal expansion and conductivity, high stiffness and strength, are the primary drivers. In this industry, performance often outweighs cost considerations in materials development.

① **Aircraft structures** MMCs have been used in several applications in aerospace components. This stems in large part to the fact that materials with increased specific stiffness and strength can significantly enhance the performance of the aircraft.

Fig. 6.8 shows an application of MMC in the fan exit-guide vane of a Pratt & Whitney engine on a Boeing 777. The MMC replaced a carbodepoxy composite that had problems with foreign object damage (FOD). The increased specific stiffness characteristics of MMCs were also exploited in a fuselage strut application considered by Airbus. The MMC replaced a carbon fiber reinforced polymer for the strut, in order to reduce cost and increase damage tolerance.

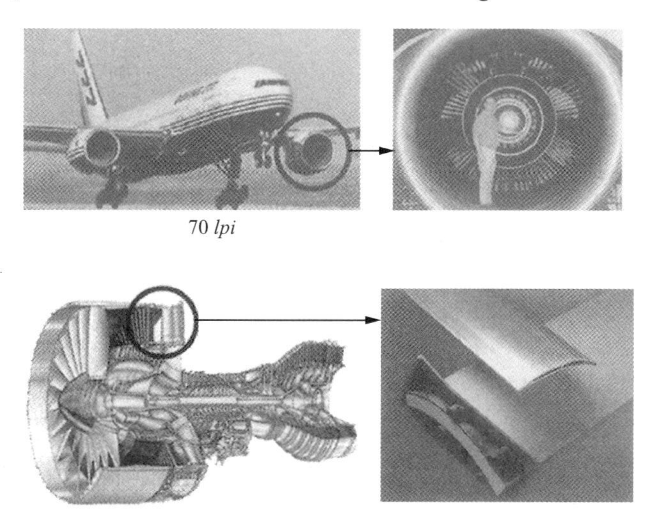

Fig. 6.8 Application of a SiC particle reinforced Al MMC in the fan exit-guide vane of a Pratt & Whitney engine on a Boeing 777. The MMC replaced a carbodepoxy composite that had problems with foreign object damage (FOD) and at a lower cost

② **Fiber metal laminates** Fiber Metal Laminates (FMLs) consist of alternately stacked thin (about a millimeter or less in thickness) sheets of metal (commonly aluminum) and fiber reinforced polymer (commonly epoxy) (Vlot and gunnick, 2001; Vogelesang et al, 1995). The first commercial FML was ARamid Aluminum Laminates, ARALL. Arall was used in a few select aircraft components, but it had structural limitations that prevented wider use. Glare or glass reinforced laminates, a glass-aluminum FML, was developed in part to overcome these limitations. Glare consists of alternating layers of aluminum and glass fiber/epoxy layers. Such a laminate is produced in an autoclave for the curing of the polymer matrix. The different layers of the laminate are stacked before curing, manually or by automated machines. There is a high degree of tailorability in these composites because both the number of layers as well as the direction of the fibers in the PMC layers can be varied depending on the application of the structural part. The main characteristics of Glare are extremely good fatigue properties, high damage tolerance capabilities, and optimal impact properties. Many structural parts (primary and secondary) can be identified where these properties play an important role. In particular, Glare has been chosen for use in the fuselage of the twin deck, 550-seat, Airbus 380 aircraft.

Fig. 6.9 shows schematically the configuration of metallic sheets and sheets of polymer matrix composite. Besides having superior fatigue behavior, glare is also lighter than aluminum, cutting the weight of the Airbus A380 by 1000 kg. It was discovered that bonded and laminated aluminum had favorable resistance to crack growth because cracks would grow in a single layer at a time, and the remaining layers would effectively bridge the crack. This discovery was made use of in the development of the F-27 aircraft wings, which is a very highly fatigue sensitive structure.

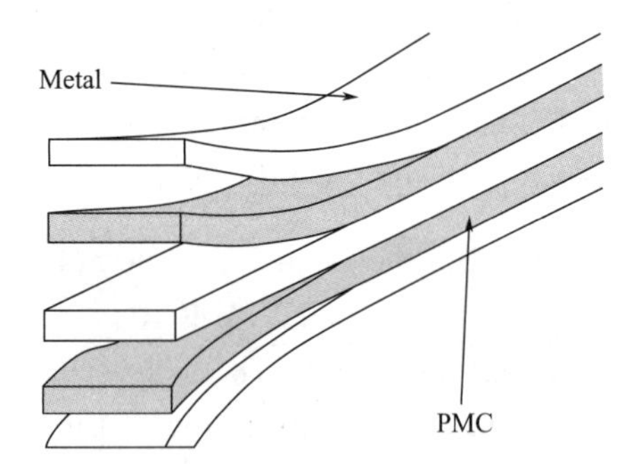

Fig. 6.9 Schematic of fiber metal laminate consisting of alternating sheets of metal and fiber reinforced polymer matrix composite

③ **Missiles** An important application for MMCs is in missiles (Shakesheff and Purdue, 1998). With increasing performance demands for missiles, conventional aluminum alloys do not have the required strength and temperature resistance. Steel and titanium are not acceptable from a weight point of view. MMCs offer enhanced strength and stiffness with no penalty in weight. In addition the elevated temperature exposure of the missile is for a very short duration (from the time it is launched to where it meets its target). Thus, MMCs are being considered in missile wings and fins, Fig. 6.10.

(2) Transportation (automotive and railway) MMCs have been used in a variety of automotive applications. An early and successful engine application was selectively reinforced aluminum pistons in the Toyota diesel engine (Donomoto et al., 1983). Another early MMC application in an automotive engine was a hybrid particulate reinforced Al matrix composite used as a cylinder liner in the Honda Prelude, Fig. 6.11. The composite consisted of an Al-Si matrix with 12% Al_2O_3 for wear resistance, and 9% carbon for lubricity. The composite was integrally cast with the engine block, had improved cooling efficiency, and exhibited improved wear and a 50% weight savings over cast iron, without increasing the engine package size. While this concept was initially implemented

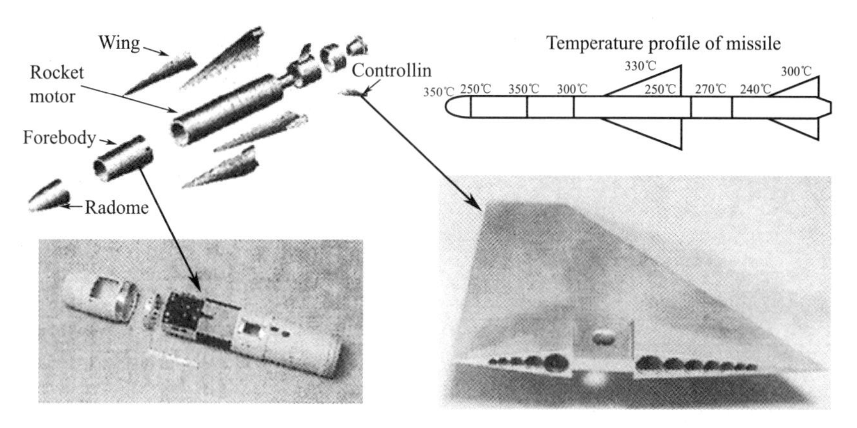

Fig. 6.10 MMC applications in missiles. The elevated temperature exposure of the missile is for a very short duration

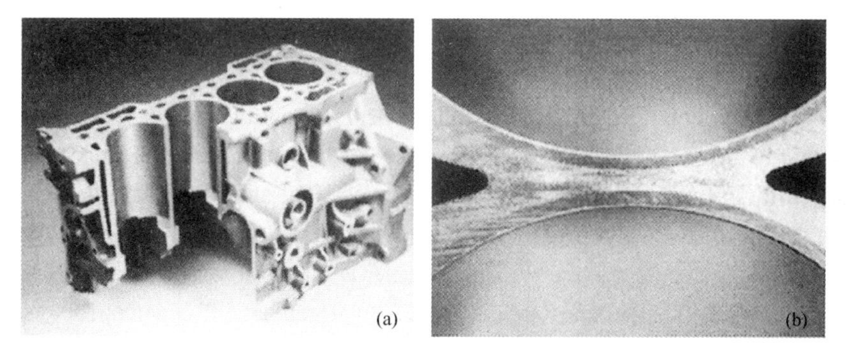

Fig. 6.11 Hybrid particulate reinforced Al matrix composite used as a cylinder liner in the Honda Prelude. The composite consisted of an Al-Si matrix with 12% Al_2O_3 for wear resistance, and 9% carbon for lubricity. (a) Prelude engine block, (b) magnified view of cylinder liner

in the Honda Prelude 2.3L engine, it has also been used in the Honda S2000, Toyota Celica, and Porsche Boxtser engines (Hunt and Miracle, 2001).

(3) Electronics and thermal management A very important market area for aluminum and to a lesser degree copper matrix MMCs is in electronic packaging and thermal management. Metal matrix composites can be tailored to have optimal thermal and physical properties to meet requirements of electronic packaging systems, e.g., cores, substrates, carriers, and housings. The main attraction of MMCs for these applications is controlled thermal expansion with a negligible penalty in thermal conductivity.

Examples of SiC particle reinforced aluminum used in microprocessor and optoelectronic packaging applications are shown in Fig. 6.12.

(4) Filamentary superconducting magnets Filamentary superconducting composites have some very important applications. Examples of applications of metal matrix composite superconducting coils include (Cyrot and Pavuna, 1992).

An important landmark in the development of high-temperature oxide superconductors occurred in 1997, when Geneva's electric utility, SIG , put an electrical transformer using HTS wires into operation. This transformer was built by ABB; it used the flexible HTS wires made by American

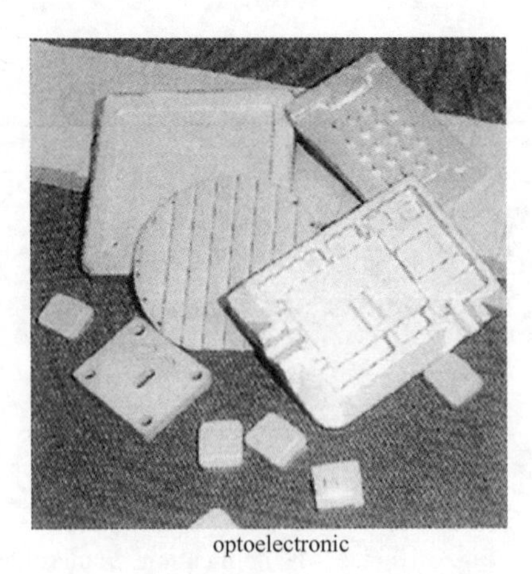

Fig. 6.12 Examples of SiC particle reinforced Al matrix composites used in microprocessor and optoelectronic packaging applications

Superconductor by the process involving packing of the raw material into hollow silver tubes, drawing into fine filaments, grouping the multifilaments in another metal jacket, further drawing and heat treating to convert the raw material into the oxide superconductor. This transformer loses only about one fifth of the AC power losses of the conventional ones. Because HTS wires can carry a higher current density, this new transformer is more compact and lighter than a conventional transformer. Liquid nitrogen used as a coolant in the HTS transformer is safer than oils used as insulators in conventional transformers. These superconductors are termed "first-generation" superconductors, Fig. 6.13(a). Newer, second-generation coated superconductors, made by the rabits process, consist of a superconductor layer, coated with a noble metal on an alloy substrate, Fig. 6.13(b).

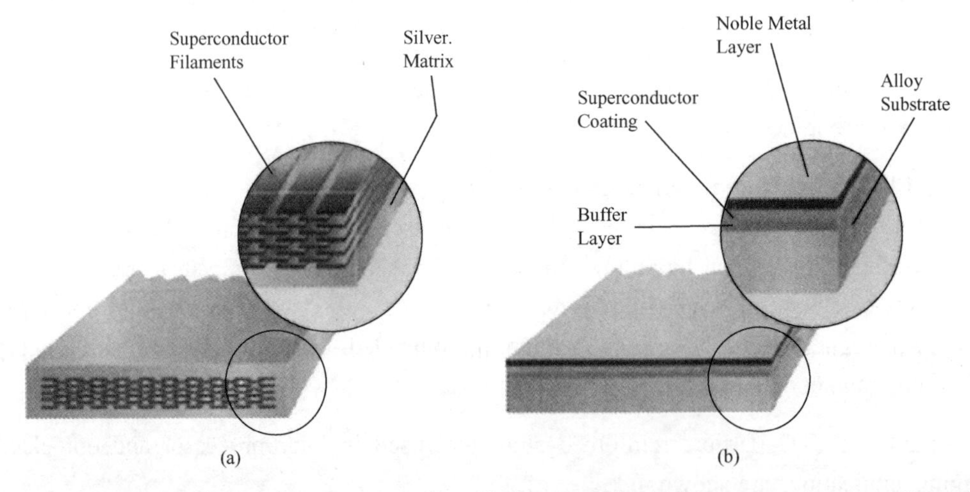

Fig. 6.13 Superconducting metal matrix composites: (a) first generation high temperature superconductor (HTS) filaments in a silver matrix (in commercial production), (b) second generation coated conductor composite, consisting of a superconductor layer, coated with a noble metal, on an alloy substrate

(5) Power conductors A relatively new application of MMCs is in the field of power transmission cable (3M, 2003). The cable consists of a composite core, consisting of Al_2O_3 continuous fibers (Nextel 610) in an Al matrix, and is wrapped with Al-Zr wires. Fig. 6.14 shows a power line with composite conductors installed in Buckeye, Arizona. The composite core bears most of the load, due to its much higher stiffness and strength. The maximum allowable temperature of the core (300℃) is also higher than that of the surrounding wires (240℃).

(6) Recreational and sporting goods Particulate metal matrix composites, especially with light metal-matrix composites such as aluminum and magnesium, also find applications in automotive and sporting goods. In this regard, the price per kg becomes the driving force for application. An excellent example involves the use of Duralcan particulate MMCs to make mountain bicycles. The Specialized Bicycle Co. in the Unites States sells these bicycles with the frame made from extruded tubes of 6061 aluminum containing about 10% alumina particles. The primary advantage is the gain in stiffness.

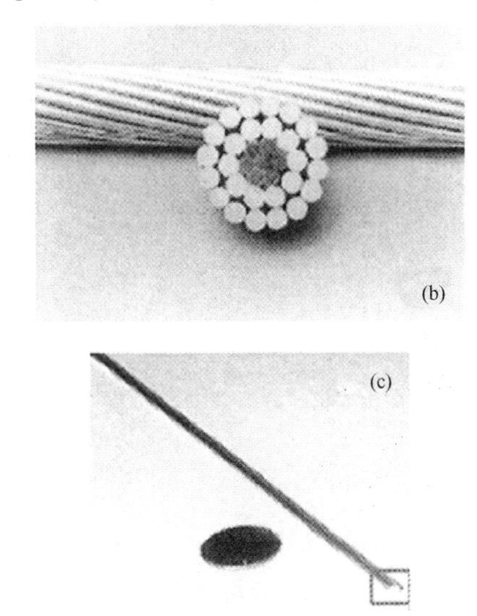

Fig. 6.14 Continuous fiber aluminum composite conductors: (a) 230 kV power line in Buckeye, AZ, (b) stranded composite conductor with aluminum alloy cables surrounding the composite core, and (c) individual composite filament

An interesting recreational application of MMCs is in track shoe spikes (Grant, 1999), shown in Fig. 6.15. The composite consists of an aluminum alloy matrix reinforced with particulate reinforcement of aluminum oxide; silicon carbide, boron carbide, or titanium carbide. The volume fraction of particles can range between 5%-30%. The composites are processed using a technique called "progressive cold forging", where about 300 parts/minute can be produced. The unique shapes of the spikes are designed to compress the track without providing unwanted impact and stress to the athlete's feet and legs.

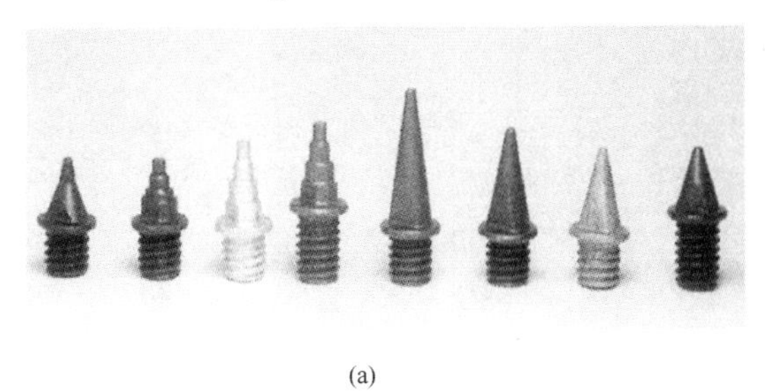

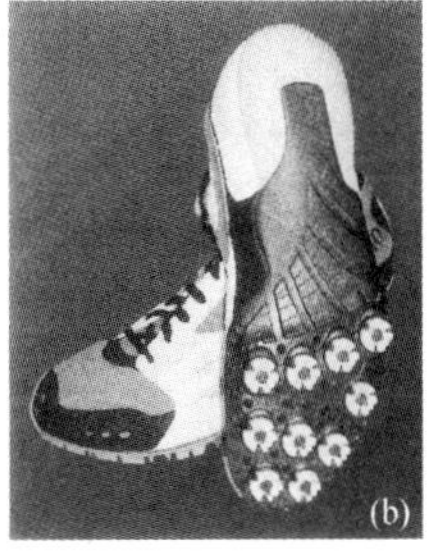

Fig. 6.15 (a) Particle reinforced MMCs used in track shoe spikes and (b) track shoe with MMC spikes

(7) Wear-resistant materials Extensive use if made of metal matrix composites in applications requiring wear resistance. Carbides, in general, and tungsten carbide, in particular, are very hard materials. Combined with a suitable ductile metal matrix we can get a composite that is very useful for cutting, grinding, drilling, roll surfaces for rolling mills and nibs of dies for wire drawing and such operations. Microstructurally, tungsten carbide based composites are isotropic and quite homogenous. They have provided a technically very effective and reliable and economically very reasonable product. Fig. 6.16 shows a roller-cone bit used for oil well drilling; the rock cutting inserts, made of WC/Co composites, on the cones.

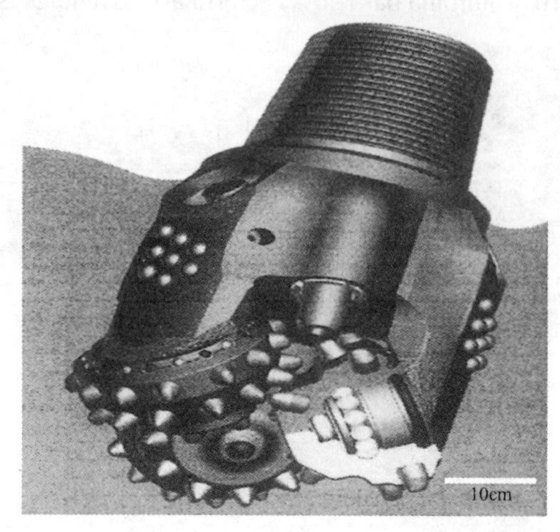

Fig. 6.16 Roller-cone bit used for oil well drilling. The rock cutting inserts are made of WC/Co metal matrix composites

New words and expressions

thermal *adj.* 热的，热量的
filamentary *adj.* 单纤维的
aircraft *n.* 飞机，航空器
recreational *adj.* 消遣的，娱乐的
laminates *n.* 层压制品
fuselage *n.* 机身（飞机）
missile *n.* 导弹
piston *n.* 活塞
lubricity *n.* 光滑
transformer *n.* 变压器
compress *vt.* 压紧
superconducting *adj.* 超导（电）的
insulator *n.* 绝缘体
automotive *adj.* 自动的，汽车的
hybrid *adj.* 混合的
extruded tube *n.* 挤压管材
ductile *adj.* 柔软的；易延展的

Notes

(1) There is a high degree of tailorability in these composites because both the number of layers as well as the direction of the fibers in the PMC layers can be varied depending on the application of the structural part.
因为复合材料的层数和聚合物基体层（PMC）中纤维取向，都能随着结构零件的用途而改变，这些复合材料存在很大程度的可修整性。

(2) Metal matrix composites can be tailored to have optimal thermal and physical properties to meet requirements of electronic packaging systems, e.g., cores, substrates, carriers, and housings.
为了满足如核、基体、载体、外罩等电子封装体系的要求，金属基复合材料可被调整为拥有最佳的热学和物理性能。

(3) It used the flexible HTS wires made by American Superconductor by the process involving packing of the raw material into hollow silver tubes, drawing into fine filaments, grouping the multifilaments in another metal jacket, further drawing and heat treating to convert the raw material into the oxide superconductor.
它采用美国超导公司生产的韧性高温超导体（HTS），具体工艺如下：将原料装满中空的银管，拉拔成单丝，并将很多单丝收集在另一种金属套管中，进一步拉拔和热处理，从而将原料转变成氧化物超导体。

(4) Combined with a suitable ductile metal matrix we can get a composite that is very useful for cutting, grinding, drilling, rollsurfaces for rolling mills and nibs of dies for wire drawing and such operations.
结合适当的韧性金属基体我们可以得到复合材料，这种复合材料对切割、打磨、钻孔、轧机、拉丝模具以及一些实际操作是非常有用的。

Exercises

1. Question for discussion

(1) What does the Fiber Metal Laminates (FMLs) consist of ?
(2) In what fields do the MMCs apply ?
(3) What are the primary drivers in aerospace applications ?
(4) What did liquid nitrogen act as in the HTS transformer ?
(5) What is the relatively new application of MMCs ?

2. Translate the following into Chinese

specific stiffness　　glass fiber epoxy layers
highly fatigue sensitive　　titanium carbide

(1) Particulate metal matrix composites, especially with light metal-matrix composites such as aluminum and magnesium, also find applications in automotive and sporting goods.
(2) There is a high degree of tailorability in these composites because both the number of layers as well as the direction of the fibers in the PMC layers can be varied depending on the application of the structural part.
(3) The composites are processed using a technique called “progressive cold forging”,where about 300 partslminute can be produced.
(4) Extensive use if made of metal matrix composites in applications requiring wear resistance.

3. Translate the following into English

各向同性材料　　丝状超导磁体
光电包装　　碳化物

(1) 纤维金属层合板（FMLs）由交替堆叠的（约 1mm 的厚度或更薄）的金属薄板（通常是铝）和纤维增强聚合物（通常是环氧）组成。
(2) 用于高温超导变压器冷却的液氮比传统变压器中绝缘用的废油更安全。
(3) 颗粒金属基复合材料，特别是轻金属（如铝、镁）基复合材料，还可以应用于汽车和体育用品等方面。

Chapter 7 Ceramics matrix composite

Unit 1 Introduction

A great variety of silicate matrices have been considered for the fabrication of fiber-reinforced glass and glass-ceramic matrix composites. Typical matrices investigated are listed in Tab. 7.1. Tab. 7.2 gives an overview of different composite systems developed and some of the most remarkable properties achieved.

Tab. 7.1 Some glass and glass-ceramic matrices commonly used to fabricate fiber-reinforced composites

Material	Elastic Modulus/GPa	Thermal Expansion Coefficient/($\times10^{-6}$/℃)	Fracture Strength/MPa	Density/(g/cm^3)	Softening Point/℃
Glass matrices:					
Silica glass	84	1.8	70~105	2.5	1300
Borosilicate	63	3.3	70~100	2.2	815
Aluminium Silicate	90	4.1	80	2.6	950
Glass-ceramic matrices:					
Lithium aluminosilicate	100	1.5	100-150	2.0	n.a.
Cordierite(magnesium aluminosilicate)	119	1.5-2.5	110-170	2.6~2.8	1450
Barium magnesium aluminosilicate	125	1.2	140	2.8	1450
Calcium aluminosilicate	110	n.a.	100-130	3.0	1550

(1) Carbon fiber-reinforced glass matrix composites The first paper on the fabrication and characterization of carbon fiber-reinforced glass matrix composites was published in 1969. Major developments in these composite systems were carried out during the 70s and 80s, especially in USA, England and Germany. There are still considerable research efforts worldwide in this area. A recent development is the use of nitride glass matrices, for example Y-Si-Al-O-N glass.

The interfacial properties of carbon fiber composites depend primarily on the physical structure and chemical bonding at the interface and on the type of carbon fiber used. Moreover, the interfacial strength in these composites can be influenced by changing the chemistry of the matrix. The major disadvantage of these composites is the limited temperature capability in oxidising atmospheres at high temperature. The oxidation behaviour under different conditions has been investigated.

(2) Nicalon and Tyranno fiber-reinforced glass and glass-ceramic matrix composites Glass and glass-ceramic matrix composites reinforced by SiC-based fiber of the type Nicalon or Tyranno combine strength and toughness with the potential for high temperature oxidation resistance. A great variety of silicate matrices have been reinforced by these fibers (see also Tab. 7.2). Several products have reached commercial exploitation, such as the material FORTADUR (Schott Glas, Germany) and the composite Tyrannohex (Ube Industries, Japan). Glass matrices such as silica, borosilicate and aluminosilicate have been used, as well as glass-ceramic matrices, such as lithium aluminosilicate (LAS), magnesium aluminosilicate (MAS), calcium aluminosilicate (CAS), barium aluminosilicate (BAS), barium magnesium aluminosilicate (BMAS), calcium magnesium aluminosilicate (CMAS), yttrium aluminosilicate (YAS), lithium magnesium aluminosilicate (LMAS) and yttrium magnesium aluminosilicate (YMAS). Oxynitride glass matrices (e.g. of $Y_{44}Si_{81}Al_{48}O_{240}N_{40}$ and $Li_{40}Si_{80}Al_{40}O_{216}N_{16}$ compositions) have also been considered. The use of refractory glass-ceramic matrices has the objective to develop materials with high temperature capability (>1000℃).

Tab. 7.2 Overview of fiber-reinforced glass and glass-ceramic matrix composites (σ: ultimate fracture strength, K_{Ic}: fracture toughness, 2-D: 2 dimensional reinforcement)

Matrix/fiber	Properties investigated
Borosilicate/SiC(Nicalon)	σ=1200MPa, K_{Ic}=18MPm$^{1/2}$(up to 600℃)
Magnesium aluminosolicate/	σ=60MPa (up to 600℃)
SiC-monofilament	σ=70MPa (up to 1100℃ in air)
Lithium aluminosolicate/SiC-(Nicalon)	σ=1100MPa (up to 1100℃ in argon)
Silica/carbon	σ=800MPa (at room temperature)
Borosilicate/metal(Ni-Si-B alloy)	σ=225MPa (up to 500℃)
Borosilicate/SiC(Nicalon)	σ=840MPa, K_{Ic}=25MPm$^{1/2}$(up to 530℃)
Aluminosilicate/SiC(Nicalon)	σ=1200MPa, K_{Ic}=36MPm$^{1/2}$(up to 700℃)
Borosilicate/SiC(Nicalon)	σ=800-1000MPa, K_{Ic}=35MPm$^{1/2}$(up to 400℃)
Borosilicate/mullite	σ=150MPa, K_{Ic}=2.5MPm$^{1/2}$(up to 600℃)
Silica/SiC(Nicalon), 2-D	σ=205MPa (up to 1000℃)
Magnesum aluminosilicate/SiC(Nicalon)	σ=1057MPa (up to 500℃)
	σ=414MPa (up to 700℃)
Barium-magnesium luminosicate/SiC(Nicalon)	σ=900MPa (up to 1100℃)

A distinct member of the family of oxycarbide fiber materials is Tyrannohex, a composite developed on the basis of bonded Tyranno fibers. The fiber content reaches values up to 90 vol.%. Due to the near absence of a matrix phase, the material retains high strength up to very high temperatures (195 MPa at 1500℃) and it has a very high creep resistance. The new generation Tyrannohex material, fabricated with novel sintered high-performance SA-Tyranno fiber has shown improved thermomechanical properties up to 1600℃.

Detailed high-resolution electron microscopy and microanalytical investigations have been conducted to elucidate the phase structure and microchemical composition of the fiber/matrix interfacial region in a variety of composite systems. These investigations showed that the interfacial zone is occupied by a carbon-rich thin layer of thickness 10-50 nm, depending on the matrix/fiber combination. The carbon-rich interfacial layer is clearly observed. High-resolution electron microscopy of these interfaces has revealed that the layers contain graphitic carbon textured to varying degrees. This carbonaceous layer is weaker than the matrix so that the fibers are effective in deflecting matrix cracks and promote fiber pull-out during composite failure. As explained below, this is the main mechanism leading to the high fracture toughness and flaw-tolerant fracture behaviour of this class of composites. At the same time, the interface layer allows load transfer from matrix to fiber so that strengthening takes place.

The mechanisms of formation of this carbon-rich layer at the fiber-matrix interfaces have been studied. It has been proposed that the carbon layer is the result of a fortunate combination of silicate matrix chemistry and non-stoichiometric/non-crystalline fiber structure. The solid-state reaction between the SiC in the fiber and oxygen from the glass and the fiber surface can be written as:

$$SiC(s) + O_2 \longrightarrow SiO_2(s) + C(s)$$

When using SiC-based fibers of the type Nicalon R and Tyranno R, the temperature capability of the composites depends strongly on the environment. In oxidising atmospheres (i.e. hot air), the stability of the carbon-rich fiber-matrix interface represents the limiting factor. One way to alleviate the problem of degradation of the carbon layer is to use protective coatings on the fibers. These coatings should act as a diffusion barrier layer and a low decohesion layer. A variety of single-layer and two-layer coatings produced by chemical vapour deposition (CVD) have been tried in attempts to replicate the mechanical response of the carbon-rich interfaces, including C, BN, BN(+C), BN/SiC and C/SiC coatings.

(3) SiC and boron monofilament-reinforced glass matrix composites Monofilament SiC fibers, produced by chemical vapour deposition (CVD), are normally fabricated in diameters ranging from 100 to 140 μm. Since these monofilaments are less flexible than the fibers derived from organo-silicon polymers, only simple shaped components can be produced. Composites were fabricated using up to 65 vol.% SiC monofilaments (type SCS-6) in a borosilicate glass matrix. The higher elastic modulus of SiC monofilament results in composites significantly stiffer than those fabricated using Nicalon or Tyranno fibers. A major disadvantage of using these filaments is their large diameter, which may lead to extensive microcracking in the matrix and, therefore, to unacceptable low off-axis strength.

SiC monofilaments have been used in borosilicate and aluminoslicate glass matrices for the fabrication of model composites, including composites with a transparent matrix. These were conveniently used to investigate the development of matrix microcracking under flexure stresses since the formation of the microcracking pattern in the transparent glass matrix could be observed in situ.

Researchers at NASA Lewis Research Centre have used SiC monofilaments for the reinforcement of glass-ceramic matrices of refractory compositions, including strontium aluminosilicate (SAS) and barium aluminosilicate (BAS), with temperature capability of up to 1600℃.

Boron monofilaments (about 20 vol.%) were used by Tredway and Prewo to reinforce borosilicate glass. They also introduced carbon fiber yarn to fill in the glass-rich regions between monofilaments in order to toughen the matrix and provide additional structural integrity of the composite.

(4) Glass/glass-ceramic fiber - glass matrix composite The first report on the preparation of this class of composites were published by Japanese researchers using *chopped* SiCaON glass fibers and matrices. The fabrication of model composites consisting of a single optical fiber embedded in borosilicate glass and of continuous oxynitride glass fiber reinforced glass matrix composites with a SiO_2-B_2O_3-La_2O_3 glass matrix has been also reported. Further work on silicate glass matrix composites with continuous silicate glass fiber reinforcement has been conducted in Germany and the UK. In a similar way as when using crystalline fibers, there is need for engineering the interface in glass/glass composites in order to avoid strong bonding which was resulted from chemical reactions during composite fabrication. A declared goal of further research in the area of glass/glass composites is the development of transparent or translucent materials showing high fracture toughness and adequate flaw tolerant behaviour. A major challenge in the development of such composites is to be able to incorporate interfaces which are optically, chemically, thermally and mechanically compatible with the matrix and fibers. One suggested way to tailor the interfaces is to use dense, transparent (or translucent) nano-sized oxide coatings. In this regard, translucent tin dioxide may be a good candidate. Other suggested oxide coating on silicate fibers used in transparent glass matrix composites is titanium dioxide produced by the sol-gel method. Glass fiber-reinforced glass matrix composite materials may lead to interesting products for replacement of laminate glass in applications requiring relatively high fracture.

(5) Metal fiber-reinforced glass matrix composites An advantage of these composites is the increased resistance to fiber damage during composite processing which results from the intrinsic ductility of metallic fibers and the possibility of exploiting their plastic deformation for composite toughness enhancement. A penalty is paid due to the relatively low thermal capability and poor chemical resistance of the metallic fiber reinforcement, limiting the application temperature and environment. In earlier study, Ducheyne and Hench fabricated composite materials with a bioactive glass (Bioglass) matrix and stainless steel fiber reinforcement by an immersion technique. Donald et al. have reported on the fabrication of glass and glass-ceramic matrix composites reinforced by stainless steel and Ni-based alloy filaments with diameters in the range of 4 to 22 μm. Glass-encapsulated metal filaments prepared by the Taylor-wire process were used for the fabrication of the composites.

Russian researchers have demonstrated the use of 2-dimensional metal fiber structures to reinforce glass and glass-ceramics. However, only a limited number of glass matrices and metallic fiber reinforcements were tried. More recently, Boccaccini and co-workers have used electrophoretic deposition to fabricate a number of glass matrix composites containing 2-dimensional metal fiber reinforcement. In particular, soda-lime, borosilicate, cathode-ray tube recycled glass and bioactive glass were used as matrices and a variety of commercially available stainless steel fiber mats were used as reinforcement. The fracture surface exhibits fiber pull-out and partial plastic deformation of the fibers, which indicates flaw-tolerant behaviour of the composite.

New words and expressions

carbon fiber-reinforced 碳纤维增强
nitride *n.* 氮化物
interfacial *adj.* 界面的
oxidising atmosphere 氧化气氛
nicalon *n.* 碳化硅
silicate matrice 硅酸盐基体
commercial exploitation 商业利用
lithium aluminosilicate 锂铝硅酸盐
magnesium *n.* 镁
calcium *n.* 钙
barium *n.* 钡
yttrium *n.* 钇
refractory glass-ceramic matrice 耐高温玻璃陶瓷基体
sinter *v.* 烧结
high-resolution electron microscopy 高分辨电子显微镜
elucidate *vt.* 阐明
pull-out 剥离
alleviate *vt.* 缓解
protective coating 保护膜
chemical vapour deposition 化学气相沉积
sol-gel 溶胶-凝胶
cathode-ray tub 阴极射线管
bioactive *adj.* 具有生物活性的

Notes

(1) A great variety of silicate matrices has been considered for the fabrication of fiber reinforced glass and glass-ceramic matrix composites.
大量不同种类的硅酸盐材料被考虑用于制备纤维增强玻璃和玻璃陶瓷基复合材料。

(2) The interfacial properties in carbon fiber composites depend primarily on the physical structure and chemical bonding at the interface and on the type of carbon fiber used.
碳纤维增强复合材料的界面性能主要取决于所使用碳纤维界面处的物理结构和化学键以及所用的纤维类型。

(3) The major disadvantage of these composites is the limited temperature capability in oxidising atmospheres at high temperatures.
这些复合材料主要缺点是在高温氧化气氛下，使用温度受到限制。

Exercises

1. Question for discussion

(1) What are the Ceramic Matrix Composites?
(2) Which matrices has been considerded for the fabrication of fiber-reinforced glass and

glass-ceramic matrix composites?

(3) What does the interfacial properties of carbon fiber composites depend primarily on?

(4) How to affect the interface strength of composite materials?

(5) What are the main contents of BMAS ceramic?

(6) What are the characteristics of the Tyrannohex fibre?

(7) What are the disadvantages of using SiC fibers filaments?

(8) Why are the fibers effective in deflecting matrix cracks and promoting fiber pull-out during composite failure?

(9) How to alleviate the problem of degradation of carbon layer in the SiC fiber-reinforced glass-ceramic matrix composites?

2. Translate the following into Chinese

(1) The interfacial strength in these composites can be influenced by changing the chemistry of the matrix.

(2) Glass and glass-ceramic matrix composites reinforced with SiC-based fiber of the type Nicalon combine strength and toughness with the potential for high temperature oxidation resistance.

(3) These coatings should act as a diffusion barrier layer and a low decohesion layer.

(4) In a similar way as when using crystalline fibers, there is need for engineering the interface in glass/glass composites in order to avoid strong bonding which results from chemical reactions during composite fabrication.

(5) This carbonaceous layer is weaker than the matrix so that the fibers are effective in deflecting matrix cracks and promote fiber pull-out during composite failure.

(6) In oxidising atmospheres (i.e. hot air), the stability of the carbon-rich fiber-matrix interface represents the limiting factor.

(7) A variety of single-layer and two-layer coatings produced by chemical vapour deposition (CVD) have been tried in attempts to replicate the mechanical response of the carbon-rich interfaces, including C, BN, BN(+C), BN/SiC and C/SiC coatings.

(8) SiC monofilaments have been used in borosilicate and aluminoslicate glass matrices for the fabrication of model composites, including composites with a transparent matrix.

(9) A major challenge in the development of such composites is to be able to incorporate interfaces which are optically, chemically, thermally and mechanically compatible with the matrix and fibers.

3. Translate the following into English

断裂韧性	石墨碳	电泳沉积
界面效应	化学气相沉积	不锈钢纤维
蠕变强度	弹性模量	极限断裂强度
相结构	塑性变形	玻璃陶瓷

4. Open topic

Do you know the similarities and differences between glass-ceramics and ceramics? Think about it, and please write about 300 words.

Unit 2 Processing of ceramic matrix composites

The easy processing in comparison to polycrystalline ceramic matrix composites is one of the outstanding attributes of glass and glass-ceramic matrix composites. This is due to the ability of glass to flow at high temperatures in a similar way to resins, which is exploited in different fabrication strategies as discussed below.

(1) Slurry infiltration and hot-pressing Hot-pressing of infiltrated fiber tapes or fabric lay-ups are the most extensively used technique to fabricate dense fiber-reinforced glass matrix composites. Fig. 7.1 shows a schematic diagram of the standard fabrication process. For proper densification, the time-temperature- pressure schedule during hot-pressing must be optimized so that the glass viscosity is low enough to permit the glass to flow into the spaces between individual fibers within the tows. The processing temperature must be chosen after taking into account the possible occurrence of crystallization of the glass matrix and the "in-situ" formation of the carbonaceous interface, as explained in Section 3, when oxycarbide fibers (e.g. Nicalon or Tyranno) are used. The pressure (of about 10-20 MPa) is usually applied after the temperature reaches the softening point of the matrix glass. In this manner almost fully dense composites can be fabricated.

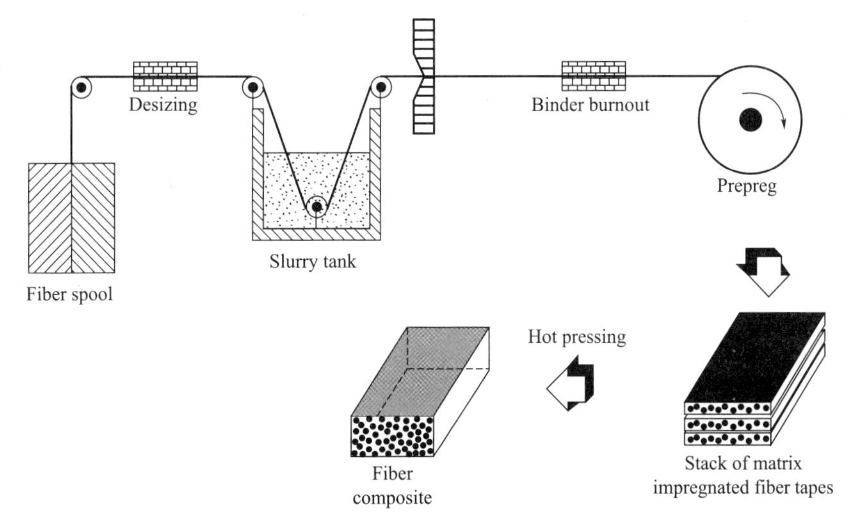

Fig. 7.1 The standard slurry impregnation and hot-pressing route to fabricate fiber-reinforced glass matrix composites

When a glass-ceramic matrix is required, the densified composites are subjected to a "ceraming" heat-treatment after densification. In this case, an optimized time-temperature "window" must be found where densification takes place by viscous flow of the glass matrix before the onset of crystallization. Ideally, the temperature range at which densification occurs at maximum rate lies between the softening temperature of the glass and the onset of crystallization temperature. A post-densification heat-treatment leads to the desired crystalline, refractory microstructure of the matrix. In some systems, the crystallization of the glass-ceramic matrix can be achieved during hot-pressing. The possibility for net shape fabrication of composite components using hot-pressing has been shown in the literature.

(2) Tape casting The equipment used for casting green sheets for use in LTCCs is shown in Fig. 7.2. Currently, a variety of casting equipments are being manufactured but in general, the equipment consists of a carrier film conveyor, casting head, slurry dispenser, drying area, and sheet take-up unit. The carrier film conveyor fulfills the role of conveying the plastic carrier film, fed from a roll, to the casting head. Since the plastic film is the carrier of the cast sheet, it is desirable that it has no wrinkles, and travels in a straight line at an even speed. At the casting head, the ceramic slurry is dispensed onto the carrier film. The slurry dispenser is for volumetric feed of the slurry to the casting head in order to produce the ceramic green sheet reliably and continuously. The drying area drives off the solvent in the cast ceramic slurry to produce a dried sheet. Drying normally uses infrared heaters or hot air. The drying temperature profile is adjusted taking into account the drying rate of the slurry and the speed of the carrier film. The sheet take-up unit picks up the dried ceramic green sheet in a roll. Some take-up units remove the green sheet from the carrier film while others take up the carrier film as well. PET (polyethylene terephthalate) film is commonly used for carrier film, and according to requirements, a silicone release agent is applied in order to improve peel-ability.

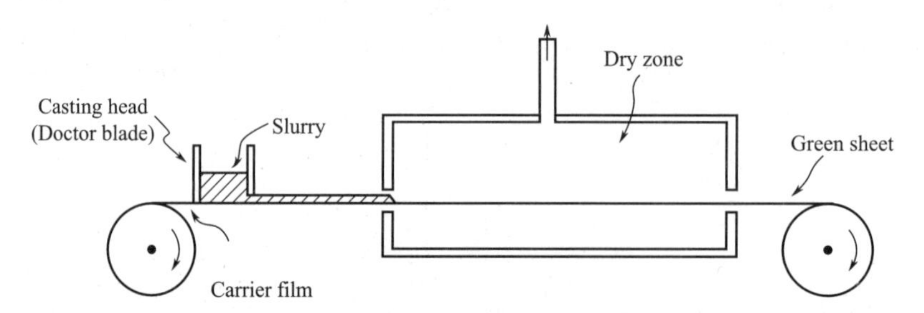

Fig. 7.2 Conceptual diagram of green sheet casting equipment

(3) Sol-gel, colloidal routes and electrophoretic deposition Using the sol-gel approach, matrices are produced from metal alkoxide solutions or colloidal sols as precursors. A great variety of glass and glass-ceramic matrices have been prepared by sol-gel processing, including borosilicate, and lithium-(LAS), magnesium-(MAS), barium-(BAS), calcium-(CAS) and sodium-(NAS) aluminosilicates. The composites are fabricated by drawing the fibers through a sol in order to deposit a gel layer on the fiber surfaces. After gelation and drying, the prepregs are densified by hot-pressing, but pressureless sintering densification is also possible. Potential advantages of the sol-gel method over slurry processing include: more effective fiber infiltration enabling a reduction of the hot-pressing temperature, which in turn leads to limitation of damage to the fibers, and the ability to tailor matrix composition in order to control thermal expansion mismatch and fiber-matrix interfacial chemistry. Another advantage of using sol-gel processing is the possibility of infiltrating complex fiber architectures and to develop nearest size and shape manufacturing technologies.

Disadvantages of the sol-gel method are associated with large matrix shrinkage during drying and densification due to low solid content of the sols, leading to matrix cracking and residual porosity. Moreover, the process is time-consuming, usually requiring a large number of infiltration/drying steps.

Another way to fabricate glass matrix composites is to combine the sol-gel with slurry approaches. This can overcome the large shrinkages inherent in the sol-gel process while maintaining the advantages of ease of infiltration and lower fabrication temperatures. An alternative process to fabricate glass and glass-ceramic matrix composites involves electrophoretic deposition (EPD) to infiltrate the fiber preform with matrix material followed by conventional pressureless sintering or hot-pressing for densification of the composite. A schematic diagram of the electrophoretic deposition cell is shown in Fig. 7.3. If the deposition electrode is replaced by a conducting fiber preform, the suspended charged (nano) particles will be attracted into and deposited within it. Using EPD, a range of glass and glass-ceramic matrix composites containing 2-dimensional fiber reinforcement, including composites of tubular shape, have been fabricated.

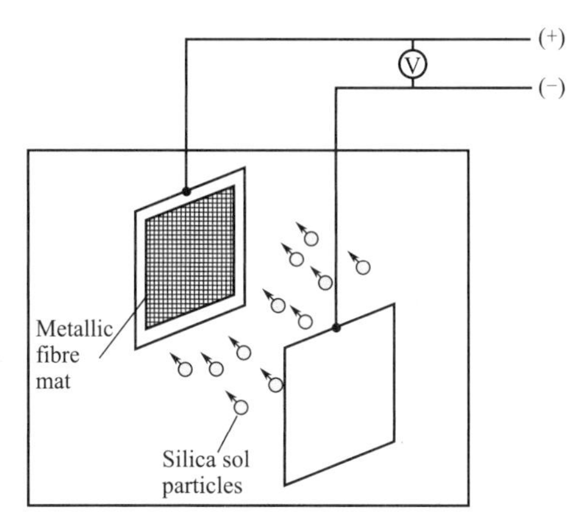

Fig. 7.3 A schematic diagram of the electrophoretic deposition cell

(4) Powder technology, sintering, hot-pressing A critical step in this fabrication approach is the mixing of the glass matrix powders and the reinforcing elements. The homogeneous distribution of these in the glass matrix is a fundamental requirement for obtaining high-quality composites with optimized properties. Inhomogeneous distribution of inclusions, forming of particle clumps or agglomerates or inclusion-inclusion interactions, may lead to microstructural defects such as pores and cracks, which will have a negative effect on the mechanical properties of the products. The improvement of the mixing techniques includes optimizing the particle sizes and the size distributions of the powders and the use of the wet-mixing routes, i.e., mixing in water or isopropanol with the addition of binders such as PVA coupled with ultrasonic or magnetic stirring. Adequate mixing of the powder matrix and the reinforcing elements is particularly problematic when chopped fibers are used, and careful control of the processing variables, including slurry viscosity, fiber content and stirring velocity is required. A fairly homogeneous glass powder-chopped fiber mixture has been achieved, which could not have been reached by dry-mixing techniques. Another approach to improve homogeneity of the mixture is the coating of the reinforcing elements by a thin layer of the matrix material. This approach presupposes the development of a technique to synthesis sub-micrometric glass particles.

The most economical process involves densification of the green bodies by simple pressureless sintering at temperatures between the glass transition and the melting temperatures of the glass matrix. Sintering in glass matrices occurs by a viscous flow mechanism. During the sintering, reactions at the inclusion/matrix interfaces or degradation of the inclusions may take place, especially when the inclusions are non-oxide ceramics (e.g. SiC) or metallic. If the glass matrix used is prone to crystallization or the aim is to produce a glass-ceramic matrix, the sintering procedure must be optimized in order to avoid the onset of crystallization before the densification by viscous flow has been fully completed. One way to achieve this is by increasing the heating rate during sintering in order to delay the nucleation and growth of crystalline phases. Thus an ideal heating schedule to produce glass-ceramic matrix composites should include three independent stages: densification, nucleation and crystallization.

The presence of inclusions can affect the formation of crystalline phases in a glass matrix. For example, it has been shown that aluminum-containing ceramic inclusions, e.g. alumina, mullite or aluminum nitride, may suppress cristobalite formation in borosilicate glass during sintering. In general, the presence of rigid inclusions will jeopardize the densification process driven by viscous flow sintering, a problem that has been well studied both experimentally and theoretically. In practice, the maximum volume fraction of inclusions suitable to yield high-quality, dense composites by pressureless sintering is about 15 vol.%. For higher contents, the consolidation of the composite powder mixtures is usually conducted by hot-pressing. This technique involves the sintering of the composite glass powder mixture under uniaxial pressure, usually in the range 5-20 MPa, in a die. While allowing the fabrication of composites with a high volume fraction of inclusions (of up to 90 vol.%) and without porosity, this technique is cost-intensive and has limitations regarding the shape and complexity of the parts that can be produced. The fairly homogeneous distribution of the reinforcing elements and the absence of porosity, cracks or other microstructural defects is evident, which indicates improved mechanical properties in these composites.

(5) Other fabrication methods A technique suggested to fabricate complex shaped structures is the matrix transfer molding technique. Woven structures used as reinforcement are arranged inside a mold cavity. Fluid matrix is transferred at high temperature into the mold cavity to fill the void space around the reinforcement structure. In this way, for example, thin-walled cylinders can be fabricated. Another proposed method for obtaining structures of complex shapes uses a super-plastically deformed foil. The foil is used to partially encapsulate the composite sample and a part holding die, which are both contained in a rigid box held in a press. This method allows near-net-shape manufacturing, but it involves complex and time consuming operations.

Efforts have also been made to evaluate the utility of the polymer precursor method for processing fiber reinforced glass-ceramics. In general, the precursor approach provides access to glass-ceramics via low temperature processing methods with good control of chemical and phase homogeneity. Because the polymer precursor method has the potential to yield near-net-shape products at relatively low temperatures, this processing route represents an interesting alternative to the established slurry and hot-pressing technique.

Another simple method of fabricating unidirectional fiber reinforced glass matrix composite rods is based on pultrusion. The technique is simple and can potentially be used for a wide range of fibers and matrices, and it enables unidirectional composites in the form of rods with different cross-sections to be fabricated. A related processing method yielding composite shapes in the form of longitudinal rods or cylinders is the extrusion technique developed by Klein and Roeder.

The densification of Nicalon fiber-reinforced borosilicate glass matrix composites by microwave heating has also been investigated. The results demonstrated that microwave processing could be a highly efficient method, saving time and reducing costs, in comparison with the traditional hot-pressing densification.

(*Selected from* Boccaccini A R.Continuous fiber Reinforced Glass and Glass-Ceramic Matrix Composites.& Roether J A.Dispersion-Reinforced Glass and Glass-Ceramic Matrix Composites.)

New words and expressions

exploit *v.* 利用，解释
hot-pressing *n.* 热压
softening point 软化点
heat-treatment 热处理
viscous flow 黏性流动
crystallization *n.* 结晶
tape casting 流延
green sheet 生料带
carrier film conveyor 承载膜传送机
casting head 流延料仓
slurry dispenser 分散浆料体系
take-up 收卷
infrared *n.* 红外线
release agent 脱模剂
peel-ability 剥离性能
electrophoretic deposition 电泳沉积法
colloidal sol 胶体溶液
precursor *n.* 前驱体
pressureless sintering 无压烧结
mismatch *vt.* 失配
infiltrating *v.* 浸润，渗透
shrinkage *n.* 收缩
solid content 固含量
residual *adj.* 残余的
encapsulate *vt.* 封装
unidirectional *adj.* 单向的

Notes

(1) Hot-pressing of infiltrated fiber tapes or fabric lay-ups is the most extensively used technique to fabricate dense fiber-reinforced glass matrix composites.
对浸润纤维带或纤维编织层进行热压是制备高密度纤维增强玻璃陶瓷基复合材料最常用的技术。

(2) The temperature range at which densification occurs at maximum rate lies between the softening temperature of the glass and the onset of crystallization temperature.
当温度范围介于软化点和结晶温度之间时，致密化速度最快。

(3) Inhomogeneous distribution of inclusions, forming of particle clumps or agglomerates or inclusion-inclusion interactions, may lead to microstructural defects such as pores and cracks, which will have a negative effect on the mechanical properties of the products.
粉料的结块、团聚、内在交联等不均匀分布情况都会造成显微结构缺陷，比如空隙和裂纹，这些都会对制品的力学性能产生负面作用。

Exercises

1. Question for discussion

(1) What does the main content of this article?

(2) Why does it say that the easy processing in comparison to polycrystalline ceramic matrix composites is one of the outstanding attributes of glass and glass-ceramic matrix composites?

(3) What is the purpose of using pressure after the temperature reaches the softening point of the matrix glass?

(4) What is the temperature range at which densification occurs at maximum rate?

(5) What are Potential advantages of the sol-gel method?

(6) What is the advantage of electrophoretic deposition over sol-gel method?

(7) Why does the glass-ceramics densification take place between the softening temperature of the glass and the onset of crystallization temperature?

(8) What are the disadvantages of the sol-gel method?

(9) Why does it say that the precursor approach provides access to glass-ceramics via low temperature processing methods?

2. Translate the following into Chinese

(1) This is due to the ability of glass to flow at high temperatures in a similar way to resins, which is exploited in different fabrication strategies as discussed below.

(2) For proper densification, the time-temperature-pressure schedule during hot-pressing must be optimized so that the glass viscosity is low enough to permit the glass to flow into the spaces between individual fibers within the tows.

(3) In this case, an optimized time-temperature "window" must be found where densification takes place by viscous flow of the glass matrix before the onset of crystallization.

(4) Disadvantages of the sol-gel method are associated with large matrix shrinkage during drying and densification due to low solid content of the sols, leading to matrix cracking and residual porosity.

(5) The homogeneous distribution of these in the glass matrix is a fundamental requirement for obtaining high-quality composites with optimized properties.

(6) Efforts have also been made to evaluate the utility of the polymer precursor method for processing fiber reinforced glass-ceramics.

(7) The densification of Nicalon fiber-reinforced borosilicate glass matrix composites by microwave heating has also been investigated.

(8) The homogeneous distribution of these in the glass matrix is a fundamental requirement for obtaining high-quality composites with optimized properties.

3. Translate the following into English

玻璃黏度	晶相	粉末烧结	均匀分布
热膨胀	金属醇盐	致密化	低温共烧陶瓷
收缩率	烧结温度		

4.Open topic

Dear students, today we are doing the experiment on the mechanical properties of ceramics. The density of ceramics is a important factor. Please think about how to improve the density of ceramics. Do you have any methods? Please, write about 200 words.

Unit 3 Glass and glass-ceramic matrix composites

The use of fiber reinforcement in silicate glass and glass-ceramic matrices started at the end of the decade after 1960. Since then, a great variety of composite systems have been developed, employing numerous matrix compositions and many types of ceramic and metallic fiber reinforcements. Interest in these inorganic composites, which constitute a particular class of ceramic matrix composites (CMCs), arises from their outstanding thermomechanical properties and potential use at temperatures up to about 1200℃. These characteristics allow their application in critical components for use in severe conditions involving relatively high temperatures, high stress levels and aggressive environments. The purpose of this article is to give an up-to-date appreciation of the underlying science and engineering of glass and glass-ceramic composite materials with fiber reinforcement, focusing on applications, processing and thermomechanical properties. The most recent literature reviews on these composite systems were published in 1995 and 2001.

(1) High-temperature, aerospace and impact resistant applications The original objective for research on fiber reinforced glass and glass-ceramic composites was to develop light materials with suitable mechanical and thermal properties for applications in gas turbines and other special areas demanding high oxidation, corrosion resistance and high temperature capability. Since these materials have approximately half the density of most Ni-based alloys, significant weight savings in components are possible. Thus, the high strength-to-weight-ratio, coupled with the high oxidation resistance, makes these composite systems candidates for general high-temperature aerospace applications, including thermal protection shrouds, leading-edges and rocket nozzle inserts. The application of fiber-reinforced glass and glass-ceramic composites in gas turbines for military and commercial propulsion has also been considered. However, cost-effective manufacturing technologies required to fabricate real component shapes (e.g. turbine blades) and complex structures have not been developed yet. For example, only limited work has been devoted to the development of machining techniques, and the joining of composite parts to ceramics and metals has also received little attention.

Another application area of fiber-reinforced glasses and glass-ceramics is in ballistic protection materials and similar impact resistant structures at room and high temperatures. In this application area, fiber reinforced-composites having higher toughness and spalling resistance than monolithic ceramics can offer a significant advantage.

(2) Automotive and special machinery applications A wide range of application possibilities exists for glass and glass-ceramic matrix composites in conventional technologies, i.e. at low to moderate temperatures and under low to moderate stresses. These applications include: components for pump manufacture (e.g. bearings and seals), automotive applications (e.g. brake and gear systems) and construction of special machinery (e.g. tools and components for use in hot environments). fiber-reinforced glass matrix composite products are being used with commercial success in the handling of hot materials, e.g. low-melting point metals and glasses, providing adequate performance in the extreme conditions of high working temperature, thermal shock loading and aggressive tribological interactions.

(3) Continuous fiber reinforced glass and glass-ceramic matrix composites Successful applications in corrosive and dusty environments, both at room and elevated temperatures, may require the development of protective coating systems. Several studies have shown that the hot corrosion of glass-ceramic matrix composites, for example under liquid sodium sulfate at 900℃ in air and argon environments, is poor: the composites show more corrosion damage than monolithic glass-ceramics. This behaviour imposes a serious limitation for uses in naval gas turbine engines where the aggressive environment of sea water and fuel will be active. Oxide coatings have been proposed to tackle this problem. Coated glass and glass-ceramic matrix composites could find application in the chemical process industry.

Glass matrix composites, in particular with carbon fiber reinforcement, have been also proposed for a variety of applications which require thermal dimensional stability, i.e. materials with multidimensional near zero thermal expansion coefficients such as support structures for laser mirrors.

(4) Electronic, biomedical and other functional applications Nicalon fiber-reinforced glass matrix composites have been proposed for producing substrates for electronic packages. The optimum thermal conductivity and dielectric constant at 1 MHz, coupled with high fracture strength and toughness, are attractive properties for this application. Low dielectric loss fiber-reinforced glass matrix composites have been also proposed for RF transmission window applications. Analysis of the literature reveals, however, that no further research has been conducted regarding functional applications of fiber-reinforced glass and glass-ceramic composites. An exploration of this research area thus remains an interesting challenge.

The biomedical field offers another potential area for broader application of composites with biocompatible glass and glass-ceramic matrices. Fiber reinforcement could be used to enhance the mechanical properties of components made of bioactive glass with the aim of fabricating load-bearing implants. Although fiber reinforcement may provide the necessary structural integrity, the biocompatibility of the composite must equally be achieved.

A futuristic application for fiber-reinforced glass matrix composites is related to the use of lunar materials for future space construction activities. Glass/glass composites in which both the fiber and the matrix are made of fused lunar soil have been proposed. These materials, obtained so far on a laboratory scale, show great promise for providing large quantities of basic structural materials for cost-effective outer-space construction.

(5) Dispersion-reinforced glass and glass-ceramic matrix composites The composites are used in their bulk form or as coatings on metallic implants. It has been shown that besides the improvement of strength and toughness, the incorporation of the ductile phase led to a significant reduction in slow crack growth. The inclusions also decrease the sensitivity of the glass to stress corrosion cracking. In related research on biomaterials, composites consisting of a dispersed alumina phase in glassy matrices are being increasingly considered as dental materials as ZrO_2, containing glass-ceramics with dispersion-type microstructure and machinable mica-containing glass-ceramics for dental restorations.

High-temperature composite coatings with high corrosion and ablation resistance are another important area of application of these materials. The potential application of tungsten particle reinforced silica in thermal protection systems, for example for the ablating surfaces of rocket nozzles and re-entry bodies, was reported 30 years ago. High-performance coatings using ceramic particle reinforced glass-ceramic matrices have been developed. Moreover, glass composite coatings consisting of SiC, carbon or intermetallic fillers in glass matrices have been developed for thermal protection of carbon/carbon components. Novel thermal barrier coatings for high-temperature components have been proposed on the basis of glass-metal (NiCoCrAlY) composites.

In enamel technology, ceramic particles can be added to improve one or several of the critical properties of enamel, such as chemical and wear resistance, impact strength and optical properties. The quality of heat resistant enamels for protection of chromium-nickel steels can thus be improved by incorporating different oxide fillers, e.g. NiO, Al_2O_3, ZrO_2, Cr_2O_3 and Fe_2O_3, into the basic silicate enamel composition.

Glass matrix composites reinforced by crystalline particles are also considered for the production of glazes with high performance. The incorporation of up to 30 vol.% of zircon ($ZrSiO_4$) particles in a transparent glassy matrix to fabricate glazes of high abrasion resistance for floor tiles has been reported.

Other application areas for partially crystallized glass-ceramics and particle reinforced glass matrix composites are those of sealing and joining of components and structures. Glass-ceramic composites containing metallic fillers are good candidates for applications as anti-friction materials. The incorporation of metallic particles increases the thermal conductivity of the composites, thus facilitating the heat removal from the contact zone within the bulk of the material. A related application area is in the handling of molten non-ferrous metals. Composite materials with high corrosion resistance to molten aluminium have been fabricated by the incorporation of Si_3N_4 particles into cordierite glass-ceramics.

Glass-ceramic materials are candidates for armour applications, including transparent armoured windscreens in vehicles and windows in buildings, as well as armour with antiballistic properties for personnel protection. Particle-reinforced glass and glass-ceramic matrix composites have potential use in this application area. The condition for transparency is either that the inclusions within the glass matrix be much smaller than the wave length of the incident light, or that the refractive index of the dispersed crystalline inclusions and the glass matrix be nearly equal. Transparent glass-ceramics and glass matrix composites are also suitable for use in electromagnetic windows, both at microwave wavelengths, and at visible and near infrared wavelengths. Other applications of transparent glass-ceramics with a matrix type of microstructure, i.e.，a glass matrix containing dispersed crystals of nanometer dimensions, have been reviewed in addition to transmission and electromagnetic.

(*Selected from* Roether J A.Dispersion-Reinforced Glass and Glass-Ceramic Matrix Composites.)

New words and expressions

silicate glass　硅酸盐玻璃
ceramic matrix composite　陶瓷基复合材料
thermomechanical property　热力学性能
gas turbine　汽轮机
thermal protection shroud　热保护罩
ballistic-protection material　防弹材料
sodium sulfate　硫酸钠
argon　*n.* 氩气
chromium-nickel steels　铬镍合金
multidimensional　*adj.* 多维的
laser mirrors　激光镜
electronic package　电子封装
dielectric loss　介电损耗
biocompatible glass　生物相容性玻璃
machinable　*adj.* 可加工的
dental restorations　牙齿修复
armour　*n.* 盔甲

Notes

(1) Interest in these inorganic composites, which constitute a particular class of ceramic matrix composites (CMCs), arises from their outstanding thermomechanical properties and potential use at temperatures up to about 1200℃.
无机化合物作为陶瓷基复合材料（CMCs）的一种特殊组分，由于它们具有显著的热力学特性和耐高温性能（可在 1200℃使用），得到了人们的极大关注。

(2) The original objective for research on fiber reinforced glass and glass-ceramic composites was to develop light materials with suitable mechanical and thermal properties for applications in gas turbines and other special areas demanding high oxidation and corrosion resistance and high temperature capability.
研究纤维增强玻璃和玻璃陶瓷复合材料的最初目的是为了开发具有合适的机械及热力学性能的轻型材料，该材料能够应用在汽轮机及其他要求高氧化、耐腐蚀和高温特性的特殊领域。

Exercises

1. Question for discussion

(1) What does the main content of this article?

(2) Which need to be considered to make glass-ceramic composite systems candidates for general high-temperature aerospace applications?
(3) What does the high-temperature composite coatings can be used to do?
(4) What was the glass-ceramic matrix composite limited for uses in naval gas turbine engines where the aggressive environment of sea water and fuel will be active?
(5) How to tachle the poor corrsion resistance of glass-ceramic matrix composites?
(6) What are the advantage of the incorporation of the ductile phase?
(7) Why can tungsten particle reinforced silica be used for high-temperature composite coatings?
(8) What can be improved by adding ceramic particles in enamel technology?
(9) Which method can be used to increases the thermal conductivity of the glass-ceramics?

2.Translate the following into Chinese

(1) The purpose of this article is to give an up-to-date appreciation of the underlying science and engineering of glass and glass-ceramic composite materials with fiber reinforcement, focusing on applications, processing and thermomechanical properties.
These characteristics allow their application in critical components for use in severe conditions involving relatively high temperatures, high stress levels and aggressive environments.
(2) The original objective for research on fiber reinforced glass and glass-ceramic composites was to develop light materials with suitable mechanical and thermal properties for applications in gas turbines and other special areas demanding high oxidation, corrosion resistance and high temperature capability.
(3) The high strength-to-weight-ratio, coupled with the high oxidation resistance, make these composite systems candidates for general high-temperature aerospace applications, including thermal protection shrouds, leading-edges and rocket nozzle inserts.
(4) The optimum thermal conductivity and dielectric constant at 1 MHz, coupled with high fracture strength and toughness, are attractive properties for this application.
(5) Fiber-reinforced glass matrix composite products are being used with commercial success in the handling of hot materials, e.g. low-melting point metals and glasses, providing adequate performance in the extreme conditions of high working temperature, thermal shock loading and aggressive tribological interactions.
(6) Only limited work has been devoted to the development of machining techniques and the joining of composite parts to ceramics and metals has also received little attention.
(7) Successful applications in corrosive and dusty environments, both at room and elevated temperatures, may require the development of protective coating systems.
(8) In this application area, fiber reinforced-composites having higher toughness and spalling resistance than monolithic ceramics can offer a significant advantage.

3.Translate the following into English

抗冲击	热膨胀系数	电子封装基板	镍基合金
防弹材料	导热系数	剥离强度	低熔点材料
介电常数	断裂强度		

4.Open topic

You have received an interview email from a electronic material company. Please reply the email to briefly introduce the application of ceramics in microelectronics field for about 200 words.

Chapter 8 Nanocomposite

扫码听音频

Unit 1 Introduction

A **nanocomposite** is as a multiphase solid material where one of the phases has one, two or three dimensions of less than 100 nanometers (nm), or structures having nano-scale repeat distances between the different phases that make up the material. In the broadest sense this definition can include porous media, colloids, gels and copolymers, but is more usually taken to mean the solid combination of a bulk matrix and nano-dimensional phase(s) differing in properties due to dissimilarities in structure and chemistry. The mechanical, electrical, thermal, optical, electrochemical, catalytic properties of the nanocomposite will differ markedly from that of the component materials. Size limits for these effects have been proposed, <5 nm for catalytic activity, <20 nm for making a hard magnetic material soft, <50 nm for refractive index changes, and <100 nm for achieving superparamagnetism, mechanical strengthening or restricting matrix dislocation movement.

Nanocomposites are found in nature, for example in the structure of the abalon shell and bone. The use of nanoparticle-rich materials long predates the understanding of the physical and chemical nature of these materials. Jose-Yacaman et al. investigated the origin of the depth of colour and the resistance to acids and bio-corrosion of Maya blue paint, attributing it to a nanoparticle mechanism. From the mid 1950s nanoscale organo-clays have been used to control flow of polymer solutions (e.g. as paint viscosifiers) or the constitution of gels (e.g. as a thickening substance in cosmetics, keeping the preparations in homogeneous form). By the 1970s polymer/clay composites were the topic of textbooks, although the term “nanocomposites” was not in common use.

In mechanical terms, nanocomposites differ from conventional composite materials due to the exceptionally high surface to volume ratio of the reinforcing phase and/or its exceptionally high aspect ratio. The reinforcing material can be made up of particles (e.g. minerals), sheets (e.g. exfoliated clay stacks) or fibres (e.g. carbon nanotubes or electrospun fibres). The area of the interface between the matrix and reinforcement phase(s) is typically an order of magnitude greater than for conventional composite materials. The matrix material properties are significantly affected in the vicinity of the reinforcement. Ajayan et al. note that with polymer nanocomposites, properties related to local chemistry, degree of thermoset cure, polymer chain mobility, polymer chain conformation, degree of polymer chain ordering or crystallinity can all vary significantly and continuously from the interface with the reinforcement into the bulk of the matrix.

This large amount of reinforcement surface area means that a relatively small amount of nanoscale reinforcement can have an observable effect on the macroscale properties of the composite. For example, adding carbon nanotubes improves the electrical and thermal conductivity. Other kinds of nanoparticulates may result in enhanced optical properties, dielectric properties, heat resistance or mechanical properties such as stiffness, strength and resistance to wear and damage. In general, the nano reinforcement is dispersed into the matrix during processing. The percentage by weight (called *mass fraction*) of the nanoparticulates introduced can remain very low (on the order of 0.5% to 5%) due to the low filler percolation threshold, especially for the most commonly used non-spherical, high aspect ratio fillers (e.g. nanometer-thin platelets, such as clays, or nanometer-diameter cylinders, such as carbon nanotubes).

(1) Ceramic-matrix nanocomposites In this group of composites the main part of the volume is occupied by a ceramic, i.e. a chemical compound from the group of oxides, nitrides, borides, silicides etc.. In most cases, ceramic-matrix nanocomposites encompass a metal as the second component. Ideally both components, the metallic one and the ceramic one, are finely dispersed in each other in order to elicit the particular nanoscopic properties. Nanocomposite from these combinations were demonstrated in improving their optical, electrical and magnetic properties as well as tribological, corrosion-resistance and other protective properties.

The binary phase diagram of the mixture should be considered in designing ceramic-metal nanocomposites and measures have to be taken to avoid a chemical reaction between both components. The last point mainly is of importance for the metallic component that may easily react with the ceramic and thereby loose its metallic character. This is not an easily obeyed constraint, because the preparation of the ceramic component generally requires high process temperatures. The most safest measure thus is to carefully choose immiscible metal and ceramic phases. A good example for such a combination is represented by the ceramic-metal composite of TiO_2 and Cu, the mixtures of which were found immiscible over large areas in the Gibbs' triangle of Cu-O-Ti.

The concept of ceramic-matrix nanocomposites was also applied to thin films that are solid layers of a few nm to some tens of μm thickness deposited upon an underlying substrate and that play an important role in the functionalization of technical surfaces. Gas flow sputtering by the hollow cathode technique turned out as a rather effective technique for the preparation of nanocomposite layers. The process operates as a vacuum-based deposition technique and is associated with high deposition rates up to some μm/s and the growth of nanoparticles in the gas phase. Nanocomposite layers in the ceramics range of composition were prepared from TiO_2 and Cu by the hollow cathode technique that showed a high mechanical hardness, small coefficients of friction and a high resistance to corrosion.

(2) Metal-matrix nanocomposites Another kind of nanocomposite is the energetic nanocomposite, generally as a hybrid sol-gel with a silica base, which, when combined with metal oxides and nano-scale aluminium powder, can form *superthermite* materials.

(3) Polymer-matrix nanocomposites definitions: In the simplest case, appropriately adding nanoparticulates to a polymer matrix can enhance its performance, often in very dramatic degree, by simply capitalizing on the nature and properties of the nanoscale filler (these materials are better described by the term ***nanofilled polymer composites***). This strategy is particularly effective in yielding high performance composites, when good dispersion of the filler is achieved and the properties of the nanoscale filler are substantially different or better than those of the matrix, for example, reinforcing a polymer matrix by much stiffer nanoparticles of ceramics, clays, or carbon nanotubes. Alternatively, the enhanced properties of high performance nanocomposites may be mainly due to the high aspect ratio and/or the high surface area of the fillers, since nanoparticulates have extremely high surface area to volume ratios when good dispersion is achieved.

Nanoscale dispersion of filler or controlled nanostructures in the composite can introduce new physical properties and novel behaviours that are absent in the unfilled matrices, effectively changing the nature of the original matrix (such composite materials can be better described by the term ***genuine nanocomposites*** or ***hybrids***). Some examples of such new properties are fire resistance or flame retardancy and accelerated biodegradability.

New words and expressions

copolymer *n.* 共聚物

fiber *n.* 纤维

carbon nanotubes *n.* 碳纳米管

exceptionally *adv.* 例外地

electrospun *n.* 电纺丝

cathode *n.* 阴极

thermite *n.* 铝热剂

tribological *adj.* 摩擦的，摩擦学的

appropriately *adv.* 适当地

biodegradability *n.* 生物降解能力

Notes

(1) A nanocomposite is as a multiphase solid material where one of the phases has one, two or three dimensions of less than 100 nanometers (nm), or structures having nano-scale repeat distances between the different phases that make up the material.
纳米复合物是一种多相固体材料，其中组成材料的不同相中至少有个在一维、二维或三维空间内尺寸小于 100 nm，或者含有纳米尺度重复单元的结构。

(2) Ajayan et al. note that with polymer nanocomposites, properties related to local chemistry, degree of thermoset cure, polymer chain mobility, polymer chain conformation, degree of polymer chain ordering or crystallinity can all vary significantly and continuously from the interface with the reinforcement into the bulk of the matrix.
Ajayan 等指出聚合物纳米复合材料的性能与其局部化学性质紧密相关，如热固塑料的固化度、聚合物链的流动性、聚合物链构象、聚合物链的有序度或者结晶度。聚合物纳米复合材料的性能从与增强体的界面到大部分基体内部都呈显著和连续变化。

(3) The percentage by weight (called mass fraction) of the nanoparticulates introduced can remain very low (on the order of 0.5% to 5%) due to the low filler percolation threshold, especially for the most commonly used non-spherical, high aspect ratio fillers (e.g. nanometer-thin platelets, such as clays, or nanometer-diameter cylinders, such as carbon nanotubes).

由于填充剂特别是通常使用的非球面、高纵横比的（如像黏土样的纳米薄层材料或者是类似碳纳米管的纳米柱状材料），具有较低的渗透域值，所提到纳米粒子的质量分数能维持在较低水平（约为 0.5% ~5%）。

(4) Definitions: In the simplest case, appropriately adding nanoparticulates to a polymer matrix can enhance its performance, often in very dramatic degree, by simply capitalizing on the nature and properties of the nanoscale filler (these materials are better described by the term nanofilled polymer composites).
定义：举个简单的例子，通过适当添加纳米粒子到聚合物基体中可显著提高它的性能，这主要是充分利用了纳米填充剂的性能（纳米填充剂应该通过“纳米填充的聚合物复合材料”来更好地表述）。

(5) Nanoscale dispersion of filler or controlled nanostructures in the composite can introduce new physical properties and novel behaviours that are absent in the unfilled matrices, effectively changing the nature of the original matrix (such composite materials can be better described by the term genuine nanocomposites or hybrids).
将填充剂或可控纳米结构分散到复合材料中能带来新的物理性能，这些新的性能是未填充的基体材料所不具备的，可以有效改变原始基体的性能（这里的复合材料应该通过“真正的纳米复合材料或者杂化材料”来更好地表述）。

Exercises

1. Question for discussion

(1) What can we learn from this passage?

(2) What forms of structure can be found as nanocomposites in nature?

(3) What field will take full use of the nanocomposites in the future?

(4) Why a relatively small amount of nanoscale reinforcement can have an observable effect on the macroscale properties of the composite?

(5) Why metallic component may easily react with the ceramic?

2. Translate the following into Chinese

coefficient of friction
restricting matrix dislocation movement
polymer chain conformation
degree of cure
the bulk of the matrix
dielectric properties
filler percolation threshold
binary phase diagram

3. Translate the following into English

软磁材料
陶瓷基复合材料
聚合物链流动性
耐腐性能
耐磨性
纵横比
真空沉积法
阻燃性

在线习题

拓展阅读

Unit 2 Processing

Organic-inorganic hybrid materials do not represent only a creative alternative to design new materials and compounds for academic research, but their improved or unusual features allow the development of innovative industrial applications. Nowadays, most of the hybrid materials that have already entered the market are synthesised and processed by using conventional soft chemistry based routes developed in the eighties. These processes are based on: ①the copolymerisation of functional organosilanes, macromonomers, and metal alkoxides; ②the encapsulation of organic components within sol-gel derived silica or metallic oxides; ③the organic functionalisation of nanofillers, nanoclays or other compounds with lamellar structures, etc.

The chemical strategies (self-assembly, nanobuilding block approaches, hybrid MOF (Metal Organic Frameworks), integrative synthesis, coupled processes, bio-inspired strategies, etc.) offered nowadays by academic research allow, through an intelligent tuned coding, the development of a new vectorial chemistry, able to direct the assembling of a large variety of structurally well defined nano-objects into complex hybrid architectures hierarchically organised in terms of structure and functions.

Independently of the types or applications, as well as the nature of the interface between organic and inorganic components, a second important feature in the tailoring of hybrid networks concerns the chemical pathways that are used to design a given hybrid material. General strategies for the synthesis of sol-gel derived hybrid materials have been already discussed in details in several reviews. The main chemical routes for all type of hybrids are schematically represented in Fig. 8.1.

(1) Path A Path A corresponds to very convenient soft chemistry based routes including conventional sol-gel chemistry, the use of specific bridged and polyfunctional precursors and hydrothermal synthesis.

Route A1: Via conventional sol-gel pathways amorphous hybrid networks are obtained through hydrolysis of organically modified metal alkoxides (vide infra section Ⅲ) or metal halides condensed with or without simple metallic alkoxides. The solvent may or may not contain a specific organic molecule, a biocomponent or polyfunctional polymers that can be cross-linkable or that can interact or be trapped within the inorganic components through a large set of fuzzy interactions (H-bonds, π-π interactions, van der Waals). These strategies are simple, low cost and yield amorphous nanocomposite hybrid materials. These materials, exhibiting infinite microstructures, can be transparent and easily shaped as films or bulks. They are generally polydisperse in size and locally heterogeneous in chemical composition. However, they are cheap, very versatile, present many interesting properties and consequently they give rise to many commercial products shaped as films, powders or monoliths. These commercial products and their field of application will be discussed in section Ⅲ-2. Better academic understanding and control of the local and semi-local structure of the hybrid materials and their degree of organization are important issues, especially if in the future tailored properties are sought. The main approaches that are used to achieve such a control of the materials structure are also schematized in Fig. 8.1.

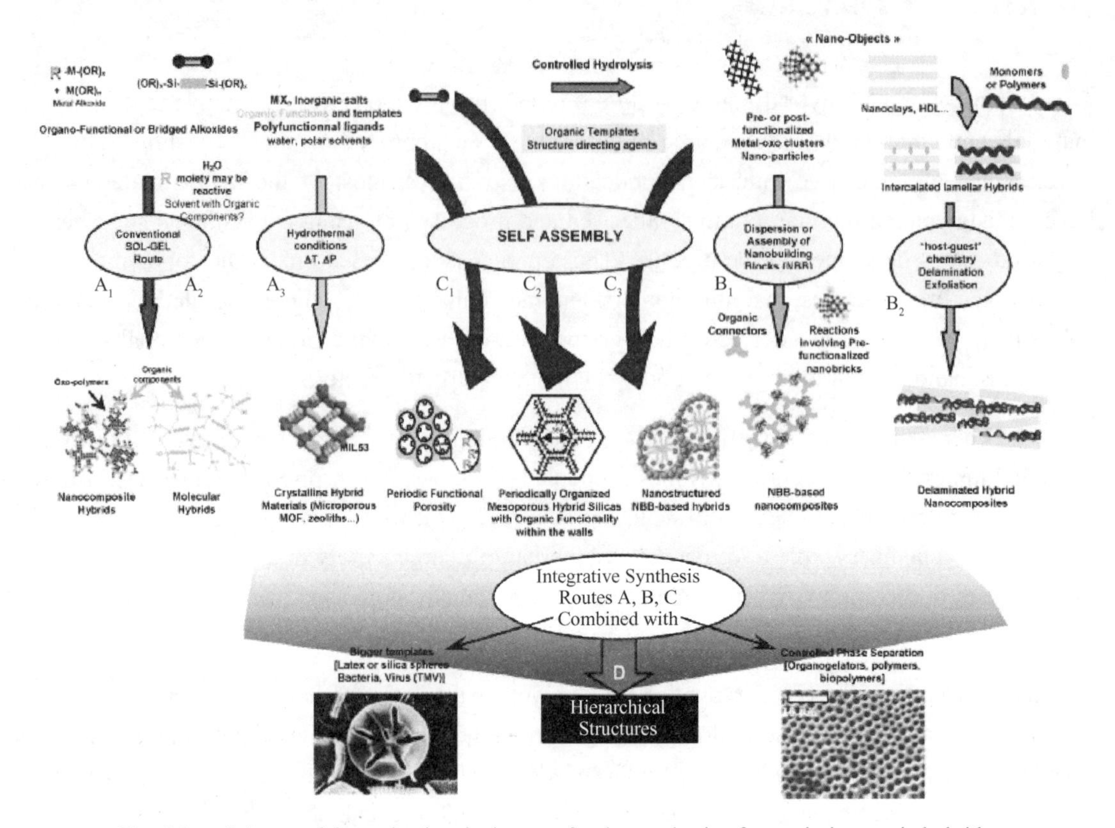

Fig. 8.1 Scheme of the main chemical routes for the synthesis of organic-inorganic hybrids

Route A2: The use of bridged precursors such as silsesquioxanes X_3Si-R′-SiX_3 (R′ is an organic spacer, X = Cl, Br, OR) allow the formation of homogeneous molecular hybrid organic-inorganic materials which have a better degree of local organisation. In recent work, the organic spacer has been complemented by using two terminal functional groups (urea type). The combination within the organic bridging component of aromatic or alkyl groups and urea groups allows better self-assembly through the capability of the organic moieties to establish both strong hydrogen bond networks and efficient packing via π-π or hydrophobic interactions.

Route A3: Hydrothermal synthesis in polar solvents (water, formamide, etc.) in the presence of organic templates had given rise to numerous zeolites with an extensive number of applications in the domain of adsorbents or catalysts. More recently a new generation of crystalline microporous hybrid solids have been discovered by several groups (Yaghi, Ferey, Cheetham and Rao). These hybrid materials exhibit very high surface areas (from 1000 to 4500 m^2/g) and present hydrogen uptakes of about 3.8 wt% at 77 K. Moreover, some of these new hybrids can also present magnetic or electronic properties. These hybrid MOF are very promising candidates for catalytic and gas adsorption based applications.

(2) Path B Path B corresponds to the assembling (route B1) or the dispersion (route B2) of well-defined nanobuilding blocks (NBB) which consists of perfectly calibrated preformed objects that keep their integrity in the final material. This is a suitable method to reach a better definition of the inorganic component. These NBB can be clusters, organically pre- or post- functionalized nanoparticles (metallic oxides, metals, chalcogenides, etc.), nano-core-shells or layered compounds (clays, layered double hydroxides, lamellar phosphates, oxides or chalcogenides) able to intercalate organic components. These NBB can be capped with polymerizable ligands or connected through organic spacers, like telechelic molecules or polymers, or functional dendrimers (Fig. 8.1). The use of highly pre-condensed species presents several advantages: they exhibit a lower reactivity towards hydrolysis or attack of nucleophilic moieties than metal alkoxides; the nanobuilding components are nanometric, monodispersed, and with better defined structures, which facilitates the characterization of the final materials.

The variety found in the nanobuilding blocks (nature, structure, and functionality) and links allows one to build an amazing range of different architectures and organic-inorganic interfaces, associated with different assembling strategies. Moreover, the step-by-step preparation of these materials usually allows for high control over their semi-local structure. One important set of the NNB based hybrid materials that are already on the market are those resulting from the intercalation, swelling, and exfoliation of nanoclays by organic polymers.

(3) Path C Path C Self assembling procedures. In the last ten years, a new field has been explored, which corresponds to the organization or the texturation of growing inorganic or hybrid networks, templated growth by organic surfactants (Fig. 8.1, Route C1). The success of this strategy is also clearly related to the ability that materials scientists have to control and tune hybrid interfaces. In this field, hybrid organic-inorganic phases are very interesting due to the versatility they demonstrate in the building of a whole continuous range of nanocomposites, from ordered dispersions of inorganic bricks in a hybrid matrix to highly controlled nanosegregation of organic polymers within inorganic matrices. In the latter case, one of the most striking examples is the synthesis of mesostructured hybrid networks. A recent strategy developed by several groups consists of the templated growth (with surfactants) of mesoporous hybrids by using bridged silsesquioxanes as precursors (Fig. 8.1, Route C2). This approach yields a new class of periodically organised mesoporous hybrid silicas with organic functionality within the walls. These nanoporous materials present a high degree of order and their mesoporosity is available for further organic functionalisation through surface grafting reactions.

Route C3: corresponds to the combination of self-assembly and NBB approaches. Strategies combining the nanobuilding block approach with the use of organic templates that selfassemble and allow one to control the assembling step are also appearing (Fig. 8.1). This combination between the "nanobuilding block approach" and "templated assembling" will have paramount importance in exploring the theme of "synthesis with construction". Indeed, they exhibit a large variety of interfaces between the organic and the inorganic components (covalent bonding, complexation, electrostatic interactions, etc.). These NBB with tunable functionalities can, through molecular recognition processes, permit the development of a new vectorial chemistry.

(4) Path D Path D Integrative synthesis (lower part of Fig. 8.1). The strategies reported above mainly offer the controlled design and assembling of hybrid materials in the 1 Å to 500 Å range. Recently, micro-molding methods have been developed, in which the use of controlled phase separation phenomena, emulsion droplets, latex beads, bacterial threads, colloidal templates or organogelators leads to controlling the shapes of complex objects in the micron scale. The combination between these strategies and those above described along paths A, B, and C allow the construction of hierarchically organized materials in terms of structure and functions. These synthesis procedures are inspired by those observed in natural systems for some hundreds of millions of years. Learning the "savoir faire" of hybrid living systems and organisms from understanding their rules and transcription modes could enable us to design and build ever more challenging and sophisticated novel hybrid materials.

New words and expressions

innovative *adj.* 革新的，创新的

derived *v.* 得到，推断；*adj.* 导出的

nanoclay *n.* 纳米黏土

hydrothermal *adj.* 热液的；*n.* 水热法

solvent *n.* 溶剂；*adj.* 有溶解力的

versatile *adj.* 通用的，用途广泛的

homogeneous *adj.* 同种的，均匀的

molecule *n.* 分子，微粒

exfoliation *n.* 剥落，剥落物

paramount *adj.* 最重要的，主要的

Notes

(1) These materials, exhibiting infinite microstructures, can be transparent and easily shaped as films or bulks.

这些纳米复合杂化材料表现出丰富的微观结构，可以做成透明的，也很容易形成薄膜或块状。transparent：透明的；film：薄膜　infinite：无限的，丰富的。

(2) These NBB can be capped with polymerizable ligands or connected through organic spacers, like telechelic molecules or polymers,or functional dendrimers.

这些小块状的纳米材料可通过聚合物配体来封顶或者用有机隔离单元来连接，例如螯合分子、聚合物、功能树状分子。NBB：nanobuilding blocks 的缩写，小块状的纳米材料；telechelic：螯合；dendrimers：树状分子。

(3) In this field, hybrid organic- inorganic phases are very interesting due to the versatility they demonstrate in the building of a whole continuous range of nanocomposites, from ordered dispersions of inorganic bricks in a hybrid matrix to highly controlled nanosegregation of organic polymers within inorganic matrices.

在自组装领域中，从无机材料在杂化基体中的有序分散，到有机聚合物在无机基体中高度可控的纳米分离，它们在构建连续纳米复合材料时所表现出多样性，使有机-无机相杂化过程非常有趣。versatility：多样性；demonstrate：展示，显示。

(4) These nanoporous materials present a high degree of order and their mesoporosity is available for further organic functionalisation through surface grafting reactions.

这些纳米多孔材料呈现出高度有序的特性，并且他们的中孔隙对通过表面接枝反应进一步有机功能化有所帮助。mesoporosity：中孔隙；available：可利用的，可得的。

(5) Learning the "savoir faire" of hybrid living systems and organisms from understanding their rules and transcription modes could enable us to design and build ever more challenging and sophisticated novel hybrid materials.
了解杂化系统和有机体系的规则和转录的"诀窍"，可以使我们能够设计和构造出越来越具挑战性和复杂的新型杂化材料。transcription：抄写，誊写；sophisticate：复杂的。

Exercises

1. Question for discussion

(1) Have most of the hybrid materials entered the market already?
(2) Which chemical strategies did academic research offer for synthesizing hybrid materials?
(3) What does the soft chemistry based routes include?
(4) What is the aim of this article?
(5) How to develop a new vectorial chemistry?

2. Translate the following into Chinese

self assembling	industrial applications
lamellar structures	metal organic frameworks
in terms of	functional dendrimers

(1) Nowadays, most of the hybrid materials that have already entered the market are synthesised and processed by using conventional soft chemistry based routes developed in the eighties.
(2) These strategies are simple, low cost and yield amorphous nanocomposite hybrid materials.
(3) These synthesis procedures are inspired by those observed in natural systems for some hundreds of millions of years.

3. Translate the following into English

化学途径	界面性质
溶胶-凝胶法	水热合成
疏水相互作用	极性溶剂

(1) 事实上，在无机与有机组分间表现出很多的界面作用（如共价键、络合作用、静电作用等）。
(2) 一体合成法是从对成千上万年自然系统的观察中所启发的。
(3) 此合成路线对应于非常简便的软化学法，包括传统的溶胶-凝胶化学合成、多功能前驱体的桥联使用，以及水热合成法。
(4) 对应于非常方便的软化学合成，包括基于传统的溶胶-凝胶化学，使用特定的桥接和多功能前体和水热合成路线。

在线习题

拓展阅读

Unit 3 Application

Looking to the future, there is no doubt that these organic-inorganic nanocomposites, born from the very fruitful activities in this research field, will open a land of promising applications in many areas: optics, electronics, ionics, mechanics, energy, environment, biology, medicine for example as membranes and separation devices, functional smart coatings, fuel and solar cells, catalysts, sensors, etc.

Commercial applications of sol-gel technology preceded the formal recognition of this technology. Likewise, successful commercial inorganic-organic hybrids have been part of manufacturing technology since the 1950s. As Arkles pointed out in a previous work, a "commercial product" is the one that is both offered for sale and used in the regular production of a device or item in general commerce. In this section, numerous examples of "commercial" as well as "potentially commercial" hybrid materials will be illustrated.

(1) Possible synergy between inorganic and organic components For a long time the properties of inorganic materials (metals, ceramics, glasses, etc.) and organic compounds (polymers, etc.) shaped as bulks, fibers or coatings have been investigated with regard to their applications, promoting the evolution of civilizations. During the last fifty years with the help of new analysis techniques and spectroscopic methods the structure/properties relationships of these materials became clearer and their general properties, tendencies and performances are well known. Some of these general properties are summarized in Tab. 8.1.

Tab. 8.1 Comparison of properties of conventional organic and inorganic components

Properties	Organics (polymers)	Inorganics (SiO_2,TMO)
Nature of bonds	Covalent[C-C](+weaker van der Waals or H bonding)	ionic or iono-covalent[M-O]
T_g(glass transition)	low(−100℃ to 200℃)	high(>200℃)
Thermal stability	low(<350℃, except polyimides, 450℃)	high(>>100℃)
Density	0.9~1.2	2.0~4.0
Refactive index	1.2~1.6	1.15~2.7
Machanical properties	elasticity	Hardness
	plasticity	Strength
	Rubbery(depending on T_g)	fragility
Hydrophobicity, permeability	hydrophilic	hydrophilic
	hydrophobic	low permeability to gases
	±permeable to gases	
Electronic properties	insulating to conductive	insulating to semiconductors (SiO_2,TMO)

Continue

Properties	Organics (polymers)	Inorganics (SiO_2,TMO)
	redox properties	redox properties(TMO)
		magnetic properties
Processability	high:	low for powders (needs to be mixed with polymers or dispersed in solutions)
	• molding casting	
	• machining	
	• thin films from solution	high for sol-gel coatings(similar to polymers)
	• control of the viscosity	

The choice of the polymers is usually guided mainly by their mechanical and thermal behavior. But, other properties such as hydrophobic/hydrophilic balance, chemical stability, bio-compatibility, optical and/or electronic properties and chemical functionalities (i.e. solvation, wettability, templating effect, etc.) have to be considered in the choice of the organic component. The organic in many cases allows also easy shaping and better processing of the materials. The inorganic components provide mechanical and thermal stability, but also new functionalities that depend on the chemical nature, the structure, the size, and crystallinity of the inorganic phase (silica, transition metal oxides, metallic phosphates, nanoclays, nanometals, metal chalcogenides). Indeed, the inorganic component can implement or improve electronic, magnetic and redox properties, density, refraction index, etc.

(2) Hybrids obtained via encapsulation of organics in sol-gel derived matrices

① **Organic molecules (dyes and "active species") in amorphous sol-gel matrices** Organic molecules play an important role in the development of optical systems: luminescent solar concentrators, dye lasers, sensors, photochromic, NLO and photovoltaic devices. However, the thermal instability of these compounds has precluded their incorporation into inorganic oxide matrices till the use of sol-gel derived glasses by Avnir, Levy and Reisfeld. Since then, many organic dyes such as rhodamines, pyranines, spyro oxazines, chromenes, diarylethenes, coumarins, NLO dyes, etc. have been incorporated into silica or aluminosilicate based matrices, giving transparent films or monoliths with good mechanical integrity and excellent optical quality, as illustrated in Fig. 8.2.

② **Bioactive sol-gel derived hybrid materials** Bioactive hybrids find applications in the field of biotechnology for the realization of biosensors and bioreactors. These fields have been recently reviewed by Livage et al. They take advantage of the high activity of enzymes, antibodies or micro-organisms to perform specific reactions that would not be possible with the usual chemical routes. Active bio-species are immobilized on or in solid substrates in order to be reusable and protect them from denaturation. Natural and synthetic polymers (polysaccharides, polyacrylamides, alginates, etc.) are currently used for bio-immobilization via covalent binding or entrapment. However inorganic materials such as sol-gel processed glasses and ceramics offer significant advantages over organic polymer hosts. They exhibit better mechanical strength together with improved chemical and thermal stability. Moreover they don't swell in most solvents preventing the leaching of entrapped bio-molecules.

Fig. 8.2 Hybrid organic-inorganic materials containing organic chromophores

Looking to the future, the control of the morphology, from the nano to the micrometer scales, associated with the incorporation of several functionalities, with a perfect control of concentrations and spatial dispersions, can yield entirely new hybrid materials with not only original and complex architectures but also optimized properties and various roles. The design and processing of innovative materials coupling original diagnostic and/or therapeutic properties such as magnetic resonance imaging (MRI) contrast, hyperthermia, radioactivity, safe transport and drugs controlled release are in progress ("FAME" network of excellence).

(3) Biomaterials and bio-inspired hybrid constructions Natural materials offer remarkable hydrodynamic, aerodynamic, wetting and adhesive properties. Evident interesting applications concern surface coatings with anti-fouling, hydrophobic, protective or adhesive characteristics and also cosmetic products.

One way to take advantage of the emerging field of biomimetics is to select ideas and inventive principles from nature and apply them in engineering products. Materials reproducing structures described in animals and plants already exist: "Riblets" are plastic films covered by microscopic grooves inspired by shark or dolphin skin that are placed on airplane wings in order to reduce the hydrodynamic trail and economize fuel; Nylon or Kevlar were inspired from natural silk and Velcro was inspired by the hooked seed of goosegrass. Concerning hybrid materials, a known example concerns super hydrophobic or super hydrophilic coatings inspired by the microstructure of lily leaves. Indeed, combining controlled surface roughness of an inorganic (glass, for example) or organic substrate in the micron-scale with hybrid layers obtained by polycondensation of hydrophobic organo-silanes yields transparent coatings with exalted hydrophobic behaviour (exhibiting wetting angles much greater than 120°, see Fig. 8.3).

(4) Class Ⅱ hybrids: ORMOCERs as an example of highly tuneable functional hybrids

① Hybrid materials for protective and decorative coatings Some interesting examples of protective ORMOCER coatings are provided by TOP GmbH, who produce lacquers developed by the Fraunhofer ISC. In fact, lacquer synthesis facility enables a large scale production of coating materials like ABRASIL, CLEANOSIL and DEKOSIL. All these lacquers are colourless and highly

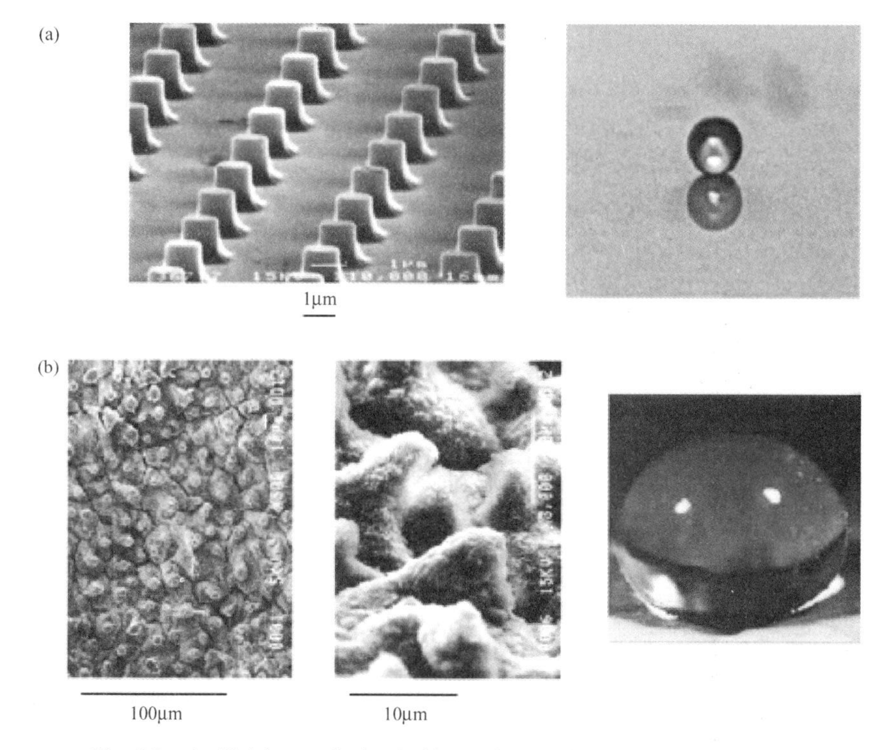

Fig. 8.3 Artificial super hydrophobic coatings (a) inspired by lily leaves (b)

transparent, which make them highly suitable for the ophthalmic market. Fig. 8.4(a) shows a plate half coated with ABRASIL and how the scratch traces from steel wool are observed on the uncoated half. This high scratch resistance is achieved with layer thicknesses of only a few micrometers. Fig.8.4(b) shows an injection moulded plastic part coated with ABRASIL. After coating and an adequate UV curing, handles are attached to make magnifying lenses for every day use.

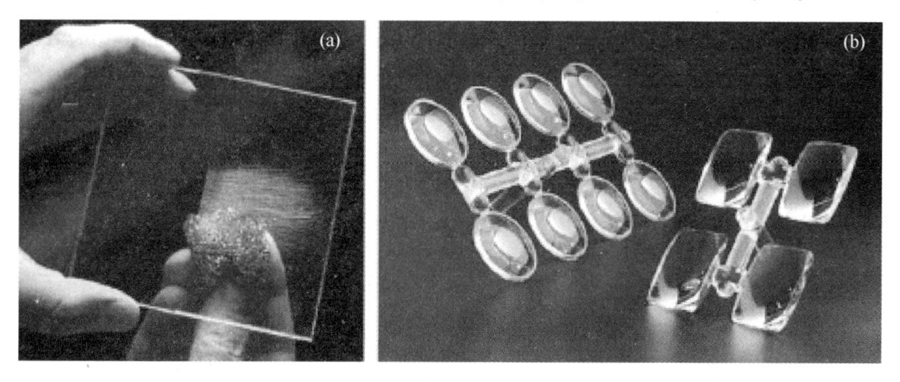

Fig. 8.4 Two examples of protective ORMOCER coatings: (a) plate half coated with ABRASIL; (b) injection moulded (plastic) part coated with ABRASIL and UV cured

② **Hybrid materials as barrier systems** The interest in hybrid materials as barrier systems has been increased in the last decades as a result of the requirements to develop much more sophisticated materials in fields such as solar cells, optics, electronics, food packaging, etc. New barrier coating materials based on ORMOCER have been developed by Fraunhofer ISC which together with a vapour-deposited SiO_x layer guarantee sufficient protection to ensure a long durability of encapsulated solar cells [see Fig. 8.5(a)]. This new inorganic-organic hybrid coating represents a whole encapsulation system since apart from the physical encapsulation it acts as an adhesive/sealing layer barrier against water vapour and gases, as well as an outside layer for weatherability. All these functions are combined in one composite ("one component encapsulant") and in this way the overall cost reduction for encapsulation reaches about 50 percent.

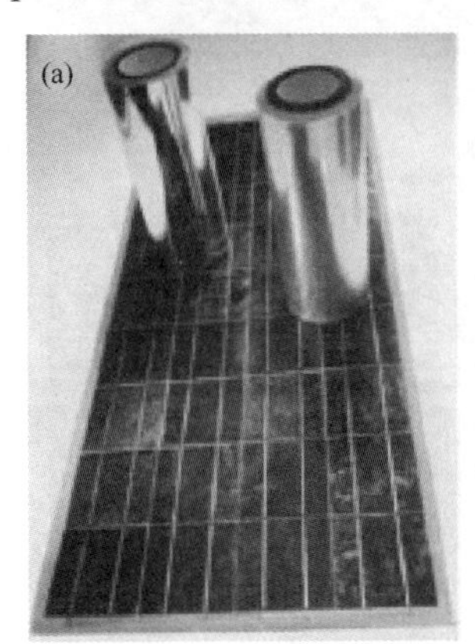

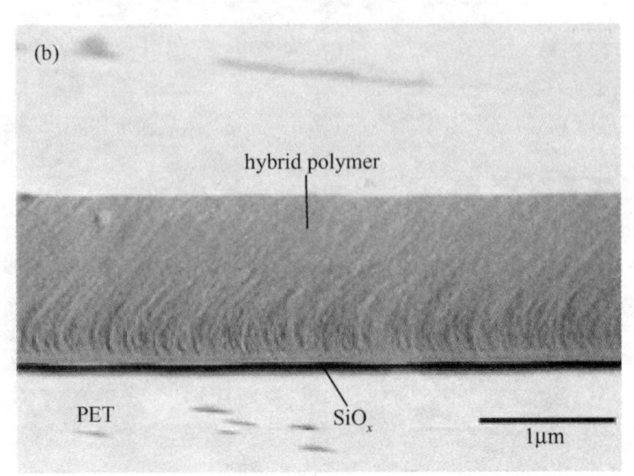

Fig.8.5 Hybrid coatings as barrier systems. (a) Aspect of the hybrid coated solar modules with very high barrier properties. (b) scanning transmission micrograph of a thin hybrid polymer coating on SiO_x deposited on a flexible PET film

The flexible nature of this hybrid material results in an optimized encapsulation process and especially in good protection of the edge area, the most difficult part for protection. Furthermore, this new material can be used from the roll, thus providing easy handling and automation in the production of flexible modules [Fig. 8.5(b)]. Thin barrier layers like SiO_x combined with ORMOCER based barrier coatings save material and energy consumption in the production of the new encapsulating material. Therefore, a material and energy saving encapsulation technology is achieved. This result should encourage the application of thin film solar cells in the building industry as well as the number of users of solar cells.

③ **Hybrid materials for dental applications** Inorganic-organic hybrid materials can be used as filling composites in dental applications. As schematized in Fig. 8.6, these composites feature tooth-like properties (appropriate hardness, elasticity and thermal expansion behaviour) and are easy to use by the dentist as they easily penetrate into the cavity and harden quickly under the effect of blue light. Moreover, these materials feature minimum shrinkage, are non-toxic and sufficiently non-transparent to X-rays. However, the composition of the hybrid material and the chemistry behind it depends strongly on its later application: as filler/particles, as matrix materials, as composites, as glass ionomer cements or as bonding.

④ **Hybrid materials for microelectronics** Organically modified resins retain important roles in electrical component coatings such as resistors and molding compounds, as well as spin-on dielectrics in microelectronic interlayer and multilayer dielectric and planarization applications. To demonstrate the feasibility of ORMOCER for use as an MCM L/D material, a substrate for a Pentium multi-chip module (MCM) with a BGA interface was realised (see Fig. 8.7).

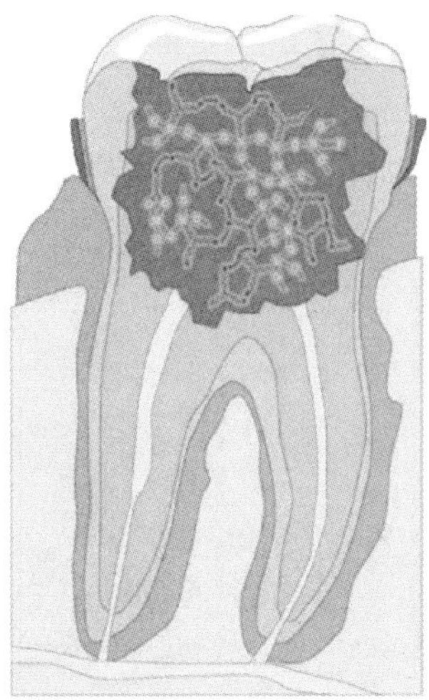

Fig. 8.6 Requirements and possibilities of dental applications of ORMOCER

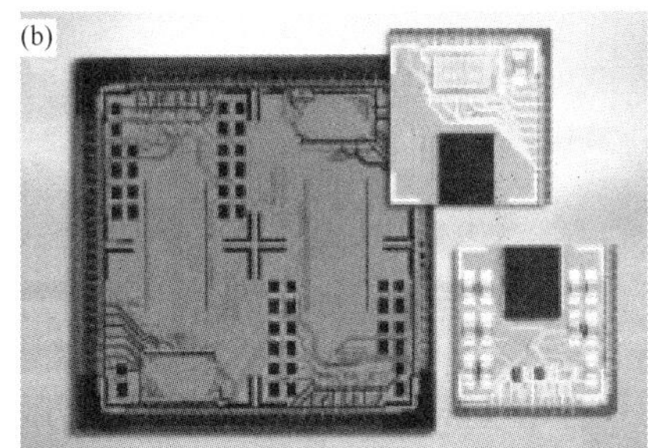

Fig. 8.7 (a) Smallest-sized Pentium MCM-L/D manufactured in ORMOCER multi-layer technology (ERICSSON/ACREO). (b) Electro-optical (o/e) MCM manufactured in ORMOCER multilayer technology (ERICSSON/ACREO/Motorola)

⑤ **Hybrid materials for micro-optics** Wafer-scale UV-embossing can be applied to substrates other than glass, for example Si and semiconductor Ⅲ-Ⅴ based wafers with prefabricated devices. In these cases, it is often advantageous to use the same hybrid materials in a combined lithographic and embossing mode to produce free-standing micro-optical elements, for example the lenslet on VCSEL elements for fiber coupling (CSEM in collaboration with Avalon Photonics Ltd., CH-Zurich). Fig. 8.8 shows SEM images of processed microoptical components on VCSEL wafers: (a) diffractive lenses, (b) an array of refractive lenses.

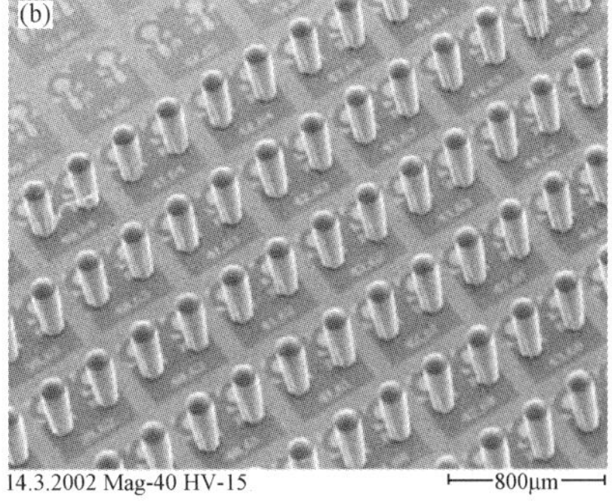

Fig. 8.8 Scanning electron micrographs of a lens array on VCSEL

New words and expressions

sol-gel technology *n.* 溶胶凝胶技术
inorganic material *n.* 无机材料，无机物质
hydrophobic *adj.* 疏水性的
implement *vt.* 实施，执行
encapsulation *n.* 封装；包装
photochromic *adj.* 光致变色的
entrapment *n.* 诱捕；截留

Notes

(1) Looking to the future, there is no doubt that these organic-inorganic nanocomposites, born from the very fruitful activities in this research field, will open a land of promising applications in many areas: optics, electronics, ionics, mechanics, energy, environment, biology, medicine for example as membranes and separation devices, functional smart coatings, fuel and solar cells, catalysts, sensors, etc.
放眼未来，毫无疑问这些在该研究领域已取得很大成果的有机-无机纳米复合材料，将在光学、电子、离子、机械、能源、环境、生物、医药等许多领域开拓更广阔的应用前景，例如有机-无机纳米复合材料可用作膜分离装置、智能薄膜、燃料和太阳能电池、催化剂、传感器等。

(2) For a long time the properties of inorganic materials (metals, ceramics, glasses, etc.) and organic compounds (polymers, etc.) shaped as bulks, fibers or coatings have been investigated with regard to their applications, promoting the evolution of civilizations.
长久以来，无论是块体、纤维或薄膜形状，无机材料（金属、陶瓷、玻璃等）和有机材料（聚合物等）的应用已经得到了广泛的研究，这极大地促进了文明的发展。

(3) The inorganic components provide mechanical and thermal stability, but also new functionalities that depend on the chemical nature, the structure, the size, and crystallinity of the inorganic phase (silica, transition metal oxides, metallic phosphates, nanoclays, nanometals, metal chalcogenides).
无机组分不但提供良好的力学及热稳定性，而且也提供一些新功能，这些新功能依赖于无机相（如硅石、过渡金属氧化物、金属磷酸盐、纳米黏土、纳米金属、金属硫化物）的化学性质、结构、尺寸和结晶度等特性。

(4) Organic molecules play an important role in the development of optical systems: luminescent solar concentrators, dye lasers, sensors, photochromic, NLO and photovoltaic devices.
有机分子在如发光的太阳能集线器，染料激光器，传感器，光致变色器，非线性光学材料和光电池装备等的光学系统发展中起着重要的角色。

(5) They take advantage of the high activity of enzymes, antibodies or micro-organisms to perform specific reactions that would not be possible with the usual chemical routes. Active biospecies are immobilized on or in solid substrates in order to be reusable and protect them from denaturation.
他们利用高活性的酶、抗体或微生物来完成常规化学法所不能完成的特殊反应。活性生物物种被固定于固体基质表面或内部，以便重复使用并防止它们变性。

Exercises

1. Question for discussion

(1) What is the sol-gel technology? Can you give some examples of its application?

(2) How can the bioactive hybrid apply in the field of biotechnology?

(3) How do you understand the word of "Riblets"?

(4) In which fields can the nanocomposite materials be applied?

(5) What's percent the overall cost can decrease by combining all functions in one composite?

2. Translate the following into Chinese

inorganic-organic hybrids   drugs controlled release
anti-fouling   bioactive hybrids
spatial dispersions   surface coatings

(1) As Arkles pointed out in a previous work, a "commercial product" is the one that is both offered for sale and used in the regular production of a device or item in general commerce.

(2) Bioactive hybrids find applications in the field of biotechnology for the realization of biosensors and bioreactors.

(3) Natural materials offer remarkable hydrodynamic, aerodynamic, wetting and adhesive properties.

3. Translate the following into English

结构   生物相容性
无机相的结晶度   折射率
光学聚合物染料   机械完整性

(1) 生物活性的杂化材料表现出了更好的机械强度以及更高的化学和热稳定性。而且它们在大多数溶剂中都不会溶胀，这可以阻止其中包裹生物分子的泄漏。

(2) 实际上将微米级的无机材料（如玻璃）或有机基体的可控表面粗糙度，与疏水有机硅烷缩聚产生的杂化层结合起来，可以生产出具有超疏水性的透明涂层。

(3) 因此，材料和节能封装技术达到了要求。此结果极大鼓励了薄膜太阳能电池在建筑行业中的应用，以及扩大了太阳能电池的用户群。

4. Scenario simulation

Supposing you are going to apply for a position in a professor's laboratory, the professor wants you to give a brief oral introduction to the application prospects and challenges of nanocomposites as biomaterials within 5 minutes.

在线习题   拓展阅读

Appendixes

Main journals of composites science and engineering

1. COMPOSITES SCIENCE AND TECHNOLOGY

   复合材料科学与技术

   ISSN: 0266-3538

2. COMPOSITES PART A-APPLIED SCIENCE AND MANUFACTURING

   复合材料 A：应用科学与制造

   ISSN: 1359-835X

3. COMPOSITES PART B-ENGINEERING

   复合材料 B：工程

   ISSN: 1359-8368

4. COMPOSITE STRUCTURES

   复合材料结构

   ISSN: 0263-8223

5. JOURNAL OF COMPOSITE MATERIALS

   复合材料杂志

   ISSN: 0021-9983

6. COMPOSITE INTERFACES

   复合材料界面

   ISSN: 0927-6440

7. APPLIED COMPOSITE MATERIALS

   应用复合材料

   ISSN: 0929-189X

8. ADVANCED COMPOSITE MATERIALS

   先进复合材料

   ISSN: 0924-3046

9. SCIENCE AND ENGINEERING OF COMPOSITE MATERIALS

   复合材料科学与工程

   ISSN: 0792-1233

10. MECHANICS OF COMPOSITE MATERIALS

    复合材料力学

    ISSN：0191-5665

Glossary

A

ablate *v.* 烧蚀
abrasion *n.*（表层）磨损处；磨损
adjacent *adj.* 邻近的，毗邻的；
adsorption *n.* 吸附
advancing angle 前进角
adverse reaction 有害反应，逆反应
aerospace *n.* 航天
aircraft *n.* 飞机，航空器
akin *adj.* 性质相同的，类似的
alleviate *vt.* 缓解
alumina *n.* 氧化铝，矾土
amorphous *adj.*无组织的，模糊的；无固定形状的，非结晶的
amorphous silica 氧化矽；非定晶硅；无定形氧化硅
amphibian *n.* 两栖动物，水旱两生植物；水陆两用车，水陆两用飞行器
adj. 两栖（类）的，水陆两用的；具有双重性的
anion *n.* 阳离子，负离子
anisotropic *adj.* 非等方向的，各向异性的
annealing *n.* 退火，韧化
anodized *adj.* 受阳极化处理的
antistatic *adj.* 抗静电的

B

bag molding 袋压成型，袋模成型
ballistic-protection material 防弹材料
balsa *n.* 西印度白塞木
barium *n.* 钡
bidimensional *adj.* 二维的
bioactive *adj.* 具有生物活性的
biocompatible glass 生物相容性玻璃
biodegradability *n.* 生物降解能力
birch *n.* 桦木
bitumen *n.* 沥青
blade *n.* 刀片；(机器上旋转的)叶片
bleach *n.* 漂白剂
boron *n.* 硼
boron filament 硼纤维
boron *n.* 硼砂
bow *n.* 箭弓
brittleness *n.* 脆性
broadgoods *n.* 宽幅
bulk *n.* 体积，本体
by-product *n.* 副产物

C

calcium *n.* 钙
capillary action 毛细管作用
carbide *n.*碳化物
carbon fiber-reinforced 碳纤维增强
carbon nanotube *n.*碳纳米管
carrier film conveyor 承载膜传送机
casing *n.* 套；罩
casting head 流延料仓
catalyst *n.* 催化剂
catastrophic *adj.* 悲惨的，灾难的，爆炸性的
category *n.*（人或事物的）类别
cathode *n.*阴极
cathode ray 电子束；电子雪崩；阴极射线
cathode-ray tub 阴极射线管
cation *n.* 阳离子，正离子
cellulosic *adj.* 纤维素的
Celsius *n.* 摄氏度 *adj.*摄氏度的
centrifugal *n.*离心
ceramic *n.* 陶瓷， *adj.* 陶瓷的，陶器的
ceramic matrix composite 陶瓷基复合材料
chalcogenide *n.*硫族化合物
chamber *n.*（作特定用途的）房间，室
chemical attack 化学腐蚀
chemical vapour deposition 化学气相沉积
Chevrolet Corvette 雪佛兰考维特，简称考维特
chopped strand mat (CSM) 短切毡
chromium-nickel steels 铬镍合金
circumferential *adj.*周向，圆周
civil infrastructure 民用基础设施
cloissone ware 景泰蓝制品
close-packed *v.* 紧密堆积；*adj.* 紧密堆积的
coir *n.* 椰子壳的纤维；棕

collagen fiber 胶原纤维
collimate *v.* 照（对；瞄）准；使成直线
colloidal silica 硅胶
colloidal sol 胶体溶液
come up with 追赶上；比得上
commercial exploitation 商业利用
compact *adj.* 小型的；紧凑的
compact bone 密质骨
compacted *adj.* 压实的，压紧的
compaction *n.* 真空压实，压实
compliance *n.* 柔度
components *n.* 部分，组件
composite *n.* 复合材料
compress *vt.* 压紧
compression *n.* 压缩，浓缩
compression molding 模压成型
compressive *adj.* 压缩的
condensation *n.* 冷凝，凝结，缩聚，缩合作用
configuration *n.* 组态，构造，结构，配置，外形
constitutive equation 本构方程
contact angle 接触角
contact molding 接触模压成型，接触模制，接触模塑
contamination *n.* 污染
continuous strand mat (CSM) 连续原丝毡
contribute *v.* 捐献，贡献，添加
control surface 控制面板
copolymer *n.* 共聚物
copolymerization *n.* 共聚合
corrugated sheet 瓦垅薄波纹板
covalent *adj.* 共有原子价的，共价的
crack *n.* 裂纹
crack advance 裂纹进展，裂纹增长
crack face 裂纹面
crack growth 裂纹扩展，裂纹增长
crack opening 裂纹张开
crack tip 裂纹端
cracking *n.* 裂纹，裂缝
creep resistance *n.* 抗蠕变性
crevice *n.* 裂缝
critical stress 临界应力
crystallinity *n.* 结晶度，结晶性
crystallization *n.* 结晶

D

damping 阻尼，衰减，内耗
debonding *n.* 脱胶，脱粘
deformation *n.* 变形，畸变
densify *v.* 压实，增浓，使增加密度
dental restorations 牙齿修复
dentine elastic modulus 质弹性模量
depression molding 减压制模
derived *v.* 得到；推断 *adj.* 导出的
dicyclopentadiene (DCPD) 二环戊二烯
dielectric loss 介电损耗
dimensional *adj.* 空间的，尺寸的
dipolar *adj.* 偶极的
disadvantage *n.* 坏处，弊病 *v.* 危害
disastrous *adj.* 灾难性的
dislocation *n.* 位错，混乱
distributed *adj.* 分布的； 分散的
drum *n.*（尤指机器上的）鼓轮，滚筒
ductile *adj.* 柔软的；易延展的
ductile fracture *n.* 韧性断裂；延性破裂
ductile iron *n.* 球墨铸铁，延性铁；韧性铁
ductile *n.* 韧性

E

ecodesign *n.* 生态概念设计[ecological 与 design 的合成词]
elastic modulus 弹性模量
electron beam 电子束，阴极射线
electronic package 电子封装
electrophoretic deposition 电泳沉积法
electroplate *vt.* 电镀 *n.*电镀物品
electrospun *n.*电纺丝
elongated cell 细长细胞
elucidate *vt.* 阐明
emerging aircraft 新式飞行器
enamel *n.* [牙]珐琅
enamel surface 釉质表面
encapsulate *vt.* 封装
encapsulation *n.* 封装；包装
entrapment *n.* 诱捕；截留
entrepreneur *n.* 企业家
epoxies *n.* 环氧基
epoxy *n.* 环氧树脂
equivalent *adj.* 相同的，同等的 *n.* 同等物
ethylene *n.* 乙烯
eutectic *adj.* 共熔的； *n.* 共晶
evangelical *n.* 新教徒; 福音派教徒; *adj.* 福音的; 新教的
exceptionally *adv.* 例外地
exfoliation *n.* 剥落；剥落物
exothermic *adj.* 放出热量的
exploit *v.* 利用，解释

extracellular matrix (ECM) 外细胞基质
extruded tube *n.* 挤压管材
extrusion *n.* 挤出，挤压

F

failure criteria 破坏（岩体、土体时应力）准则、失效判据
failure strain 失效应变
failure-inducing surface crack 缺陷诱导的表面裂纹
fatigue *n.* 疲劳
fatigue-fail 疲劳失效
feedstock *n.* 给料；原料
fiber *n.* 纤维
fiber-reinforced polymer (FRP) 纤维增强聚合物
fibrous *adj.* 纤维的; 纤维构成的; 纤维状的
filament *n.* 细丝；丝状物
filament winding process 长丝卷绕工艺（缠绕成型法）
filamentary *adj.* 单纤维的
filaments *n.* 丝状物
finite elements model (FEM) 有限元方法
flammability of polymer 聚合物阻燃性能
flank *n.* 侧面, 胁, 侧腹，翼
flaw *n.* 缺陷，裂纹
flax *n.* 亚麻
flexible *adj.* 韧性的，挠性的
flexural stiffness-to-weight ratio 弯曲刚度与重量比
flexure *n.* 弯曲；拐度，曲率；弯曲部分；单斜挠褶
flight-worthy 航空标准
floor beam 底部横梁（地楞横梁）
forging *n.* 锻造
fracture mechanics 断裂力学
fracture *n.* 断裂，断裂面
friction *n.* 摩擦
fumed silica [材]煅制氧化硅
fused silica 熔融石英；[化]石英玻璃
fuselage *n.*机身（飞机）
FVF(fiber volume fractures) 纤维体积含量

G

gas turbine 汽轮机
gelation *n.* 凝胶化（作用）；凝胶的形成（过程）
geometric *adj.* 几何学的，几何装饰的；成几何级数增减的
geometry *n.* 几何学
gold thread embroidery 金线刺绣
golf shaft 高尔夫球杆
goniometer *n.* 角度计，测角器
gradient *n.* 梯度，陡度；*adj.* 倾斜的
grain boundary 晶界
granulated *adj.* 颗粒状的
graphite *n.* 石墨，石墨纤维
green sheet 生料带

H

hallmark *n.* 特点，标志；检验印记
hand lay-up 手糊工艺
hang-glider 滑翔机
heat of formation 生成热
heat-treatment 热处理
heterogeneous *adj.* 多种多样的；不均匀的，异质的
hexagonal *adj.* 六角形的，六边形的
high-resolution electron microscopy 高分辨电子显微镜
homogeneous *adj.* 同性质的，同类的，均匀的；
horizontal stabilizer 水平尾翼
hot-pressing *n.* 热压
hybrid *adj.* 混合的
hydraulic cement mortar 液压水泥砂浆
hydraulic press *n.* 水压机
hydrophobic *adj.* 疏水性的
hydrothermal *adj.* 热液的；*n.* 水热法
hydroxyapatite 羟基磷灰石
hysteresis *n.* 滞后

I

impact *n.* 冲击
imperfection *n.* 缺陷，不足之处，不完整性
implement *vt.* 实施，执行
implement plan 执行计划，实施计划
impregnate *v.* 使充满；使遍布；浸渍
impregnated *adj.* 浸渍的
impregnation *n.* 浸胶
impurity *n.* 杂质
in compliance with 顺从；如
in lieu of 代替
inclusion *n.* 掺杂物，杂质 *v.*夹杂
incorporate *v.* 将…包括在内；包含
incorporated *adj.* 合成一体的
infiltrating *v.* 浸润，渗透
infiltration *n.* 渗透；渗滤
infinitesimally *adv.* 无限小地
infrared *n.* 红外线
ingots *n.* 钢锭，铸块
ingredient *n.* 成分；原料；
inhibitor *n.* 抑制剂

inhomogeneity *n.* 非均质性，不均匀性
initiator *n.* 引发剂
innovative *adj.* 革新的，创新的
inorganic material *n.* 无机材料，无机物质
insulator *n.* 绝缘体
interface *n.* 界面
interfacial *adj.* 界面的
interfacial zone 界面区
interlaminar *adj.* 层间
interlocking *n.* 咬合
internal stress 内应力
interrogate *v.* 质问，审问，讯问
interstice *n.* 间隙，空隙，小缝
isotropic *adj.* 各向同性的

J

jute *n.* 黄麻

K

kaolin *n.* 高岭土

L

lamina *n.* 薄板，薄层；名词复数：laminae，laminas
laminate *v.* 制成薄板，制成箔；形成薄板
adj. 由薄片组成的；薄板状的
laminated *adj.* 层压的，薄板状的
laminates *n.* 层压制品
lamination theory 层合板理论
laser mirrors 激光镜
lattice *n.* 格子金属架，晶格，点阵
lattice vibration 晶格振动
legitimate *adj.* 合理的
lignin *n.* 木质素
lithium aluminosilicate 锂铝硅酸盐
load-bearing 支撑结构
local stress 局部应力
longevity *n.* 长寿，寿命
longitudinal compressive load 纵向压载
loop *n.* 环，环路，活套
lubricant *n.* 润滑剂；润滑油
lubricity *n.* 光滑

M

machinable *adj.* 可加工的
macromechanic *n.* 宏观机理
macromolecule *n.* 大分子
magnesium *n.* 镁
mammalian bone 哺乳动物骨质
mandrel *n.* 芯轴，芯模，芯管，芯棒
manufacturing process 制造工艺
material creep 材料蠕变
matrix algebra 矩阵代数
maximize *vt.* 使（某事物）增至最大限度，最大限度地利用（某事物）
mesophase *n.* 中间相
metal matrix *n.* 金属基
metallic phosphate *n.* 金属磷酸盐
metallurgical *adj.* 冶金的，冶金学的，冶金术的
mica *n.* 云母
micromechanic *n.* 微观机理
micromechanics model 细观力学模型
minutest *adj.* 微小的
mismatch *vt.* 失配
missile *n.* 导弹
modulus *n.* 模量
mold *n.* 模子；*vt.* 浇铸
molding process 成型加工
molecule *n.* 分子；微粒
molten *adj.* 熔化的；熔融的
monolithic *adj.* 整体的
monolithic material 整体材料
monotectic *n.* 偏共晶，偏晶体
morphogenesis *n.* 形态发生，形态生成
moulding *n.* 模塑
mullite *n.* 莫来石
multidimensional *adj.* 多维的

N

nanoclay *n.* 纳米黏土
necking *n.* 颈缩
netting analysis 网格分析
neutrality *n.* 中性
nicalon *n.* 碳化硅
nitride *n.* 氮化物
nondirectional *adj.* 无方向的，非定向的，适合个方向的
non-resorbable 非吸收性缝合线
nontacky *adj.* 非黏性的，不黏的
normal stress 正应力，法向应力

O

oar for rowing 赛艇运动使用桨，橹
odontoblast *n.* 牙质
on-axis unidirectional ply 轴向单层
optimisation *n.* 最佳化、最优化；优选法
optimum *adj.* 最适，最优
orientable *adj.* 可定向的
orientation *n.* 定向，方位
orifices *n.* 孔
orthogonal lattice 直角点阵
orthotropic *adj.* [力]正交各向异性；[植]直生的，正交的
osteocyte *n.* 骨细胞
oxidising atmosphere 氧化气氛

P

parallel *adj.* 平行的
paramount *adj.* 最重要的，主要的
peel-ability 剥离性能
perpendicular *adj.* 垂直的；成直角的
phenolic *n.* 酚醛树脂，酚醛塑料
phenolics *n.* 酚醛塑料
philosophical paradigm 哲学范式
photochromic *adj.* 光致变色的
piezoelectric property 压电性能
piston *n.* 活塞
pitch *n.* 沥青
platinum *n.* 铂
plethora *n.* [医]多血，多血症
ply (layer or lamina) （毛线、绳等的）股；层；厚；（夹板的）层片
polyester *n.* 聚酯，涤纶
polyester resin 聚酯树脂
polyethylene *n.* 聚乙烯
polymeric *adj.* 聚合的
polymerization *n.* 聚合反应，聚合作用
polymethyl methacrylate 聚甲基丙烯酸甲酯
polystyrene *n.* 聚苯乙烯
poly-ureaformaldehyde 聚脲甲醛
polyurethane *n.* 聚氨酯（树脂），聚氨基甲酸树脂
polyvinyl chloride 聚氯乙烯
porous *adj.* 多孔的
pottery *n.* 陶器
precompounded *adj.* 预制的
precursor *n.* 前驱体
predetermined *adj.* 预先确定的； 预先决定的
predominantly *adv.* 占主导地，占优势地，主要地
preimpregnate *n.* 预浸（渍）
premature *adj.* 未成熟的；过早的
prepreg *n.* 预浸料坯，预浸渍制品
pressureless sinterin 无压烧结
propulsion nozzle *n.* 推进喷嘴
protective coating 保护膜
pteranodon *n.* 无齿翼龙
pterosaur *n.* 翼龙
p-tert-butylcatechol 对叔丁基邻苯二酚
pull-out 剥离
pullout *n.* 拉脱
pultrusion *n.* 拉挤成型
pure shear 纯剪切力

Q

quench *n.* 淬火，冷浸

R

radome *n.* 雷达天线罩
receding angle 后退角
recreational *adj.* 消遣的；娱乐的
refractory *n.* 耐火物质
refractory glass-ceramic matrice 耐高温玻璃陶瓷基体
reinforcement *n.* 增强体
release agent 脱模剂
residual *adj.* 残余的
residual stress 残余应力
resin *n.* 树脂
rocket motor case 火箭发动机壳体
roller *n.* 滚筒，滚轴
roughness *n.* 粗糙度
rudder *n.* 舵
rugosity *n.* 有皱纹，多皱纹性质
rust *n.* 生锈 *v.* 生锈

S

screw *n.* 螺杆，螺钉
seismic resistance 抗震
sensitivity *n.* 敏感
sessile *adj.* 静止的，不动的
shear lag 剪滞
shear strength 剪切强度
shear stress 剪应力
shelf life *n.* 保质期
shellac *n.* 虫胶
shrinkage *n.* 收缩
silica *n.* 二氧化硅；硅土
silica brick 硅砖；石英砖
silica flour [冶金]矽砂粉
silica gel *n.* [化]硅胶
silica glass 石英玻璃；硅玻璃（等于 fused silica）
silica sand 硅砂；石英砂
silica sol 硅溶胶
silica-based *adj.* 硅基的，石英基的
silicate glass 硅酸盐玻璃
silicate matrice 硅酸盐基体
silicon carbide *n.* 金刚砂，碳化硅
silicon dioxide, silicon oxide *n.* [无化]二氧化硅；[材]硅土
simultaneously *adv.* 同时地
sinter *v.* 烧结
sinusoidally *adv.* 正弦型
sisal *n.* 剑麻
slide *v.* 滑动，滑行, 滑道，滑梯
slip systems 滑移系
slurry dispenser 分散浆料体系
sodium sulfate 硫酸钠
soft, tender *adj.* [材] 柔软的，易延展的
softening point 软化点
sol-gel 溶胶-凝胶
sol－gel technology *n.* 溶胶凝胶技术
solid content 固含量
solid free-form fabrication (SFF) 实体自由成形制造
solid solution 固溶体，固体溶液
solidification *n.* 凝固,固化
solvent *n.* 溶剂；*adj.* 有溶解力的
sophisticated *adj.*复杂的，尖端的，高级的
specimen *n.* 试样，试件，样品
spongy bone 海绵骨
spontaneous *adj.* 自发的
spruce *n.* 云杉
squash racket 软式网球
squeeze out 挤出，榨出
stamp *v.* 冲压
static friction 静态摩擦力
stiffness *n.* 刚度
stoichiometric *adj.* 化学计量的
stoichiometry *n.* 化学计量学，化学计量关系
stone implement 石器；石器用具
strength *n.* 强度
strength-to-weight ratio *n.* 比强度
stress *n.*应力
subdivide *vt.* 细分
subscript *n.* 下标
subset *n.* 子集
superconducting *adj.* 超导(电)的
supercool *v.* 使过度冷却
superlattic *n.* 超晶格，超点阵
symmetric *adj.* 对称性的，均衡的
synonymous *adj.* 同义字的，同义的，类义字的

T

talc *n.* 滑石
tape casting 流延
tensile *adj.* 拉长的，拉力的，拉伸的
tensile strength 拉伸强度
terminology *n.* 术语
the Book of Exodus 出埃及记
theoretical cohesive strength 理论结合强度
thermal *adj.* 热的，热量的
thermal protection shroud 热保护罩
thermite *n.* 铝热剂
thermomechanical property 热力学性能
thermoplastic *n.* 热塑性
thermoplastic resin 热塑性树脂
thermoset *n.* 热固性
thermoset resin 热固性树脂
thermosetting *adj.* 热固性的，热凝性的，热成形的
thickener *n.* 增稠剂
tilt *n.* 倾斜，倾侧
toughness *n.* 韧性，强硬
tow *n.* 纤维束，丝束
trabecular bone 小梁骨
transformer *n.* 变压器
transmission *n.* 播送，传送
transverse *adj.* 横向的
transverse mechanical property 横向力学性能
trauma *n.* 创伤；损伤
tribological *adj.* 摩擦学的

U

ultraviolet *n.* 紫外线；*adj.* 紫外线的；*v.* 紫外线辐射
undissociable *adj.* 不可分离的
uniaxial strain 单轴应变
uniaxial tensile 单轴拉伸
uniaxially *adv.* 单向地
unidimensional *adj.* 线性的，一维的，一度空间的
unidirectional *adj.* 单向的，单向性的
uniformity *n.* 均质性
untwist *v.* 解开；未扭曲

V

valence *n.* 化合价
vapor deposition *n.* 蒸镀，气相沉积
vector *n.* 矢量
veneer *n.* 表层饰板
versatile *adj.* 通用的，用途广泛的
vis a vis 与...相比较，对于，同…相比
viscosity *n.* 黏性；黏质
viscous flow 黏性流动
void *n.* 空隙

W

waxed paper *n.*（包装食品或烹饪用的）蜡纸
wet layup 湿铺工艺
wettability *n.* 润湿性
whisker reinforced *adj.* 晶须增强的
wollastonite *n.* 硅酸钙岩矿
workability *n.* 可使用性；施工性能；可加工性
woven fabric 纺织品
woven *n.* 交叉织状；*v.* 编，织，织成（weave 的过去分词）

X

X-ray X 射线

Y

yttrium *n.* 钇